智能儿童玩具设计

鲁艺 黄韬 著

中国商务出版社

图书在版编目（CIP）数据

智能儿童玩具设计：英文 / 鲁艺，黄韬著. — 北京：中国商务出版社，2022.10（2023.8重印）

ISBN 978-7-5103-3876-2

Ⅰ. ①智… Ⅱ. ①鲁… ②黄… Ⅲ. ①智能控制—玩具—设计—英文 Ⅳ. ①TS958.02

中国版本图书馆 CIP 数据核字（2021）第 137564 号

智能儿童玩具设计
ZHINENG ERTONG WANJU SHEJI

鲁 艺 黄 韬 著

出　　版：	中国商务出版社
地　　址：	北京市东城区安外东后巷28号　　邮　编：100710
责任部门：	发展事业部（010-64218072）
责任编辑：	陈红雷
直销客服：	010-64515210
总 发 行：	中国商务出版社发行部　（010-64208388　64515150）
网购零售：	中国商务出版社淘宝店　（010-64286917）
网　　址：	http://www.cctpress.com
网　　店：	https://shop162373850.taobao.com
邮　　箱：	295402859@qq.com
排　　版：	北京凯能特教育科技有限公司
印　　刷：	河北赛文印刷有限公司
开　　本：	710毫米×1000毫米　　1/16
印　　张：	9　　　　　　　　　　　字　数：132千字
版　　次：	2022年10月第1版　　　　 印　次：2023年 8月第2次印刷
书　　号：	ISBN 978-7-5103-3876-2
定　　价：	58.00元

凡所购本版图书如有印装质量问题，请与本社印制部联系（电话：010-64248236）

版权所有　盗版必究　（盗版侵权举报可发邮件到本社邮箱：cctp@cctpress.com）

Preface

Child-oriented tangible interaction is a frontier field of the new generation of human-computer interaction, and children's products are a huge and special research field in tangible interaction systems. As the "indigenous people" of the information society, modern children have been immersed in an informationized and intelligent digital world ever since their birth. At present, the field of tangible interaction research mainly focuses on how to design intelligent interactive products that conform to children's cognitive models, have high usability, and are with edutainment features. Starting from the study of child users' psychological cognitive factors and game behavior characteristics, this book aims to explore the design method of children's products based on tangible interaction, propose a theoretical model of the design of children's products based on tangible interaction, and apply the theory to teaching practice to further verify the feasibility of the method.

This book first takes pre-school children aged 3-7 as an example, using methods such as literature review, case analysis, multidisciplinary research, and theoretical verification. Firstly, it starts from the connotation of tangible interaction theory and introduces children's cognitive psychology, children's pedagogy, and children's product innovation design methodology. Based on this research idea, this book first starts from studying the physiology, psychology, gaming behavior, emotion and other need levels of children of different ages, and uses qualitative and quantitative methods to collect and analyze a large amount of user data, and build user persona models. Secondly, by analyzing existing interaction models suitable for physical user interfaces and on the basis of the principles of interaction design in the design of children's products, the author proposes the design of children's products model and product design process based on tangible interaction. Thirdly, the book introduces the design model into teaching practice, and summarizes the future-oriented intelligent design concept and theoretical basis of children's products through a large number of design schemes and usability evaluation.

In summary, this book builds a domestic child development research data platform—"Kidsplay" by collecting and researching children's big data. This data research platform comprehensively collects and analyzes the characteristic data of Chinese children from such perspectives as physical development, psychological development, cognitive development, game development, etc., to deliver the value of big data to the traditional children industry; at the same time, it builds a child product design model based on tangible interaction in three dimensions, respectively user, interaction, and product. On that basis, it further transforms the theoretical model into a plan of children's smart product design based on tangible interaction. The software and hardware platforms of smart products meet the edutainment needs of children and parents, and further verify the feasibility of the theoretical interaction model in the design of children's products.

Contents

Chapter 1 Introduction ·· 1

    1.1 The background of research on the design of children's products ············ 2

    1.2 Domestic and foreign cases ·· 7

    1.3 Research content and research methods ··· 10

Chapter 2 Theory of the Interaction Design of Children's Products ············ 14

    2.1 Introduction ·· 15

    2.2 Theoretical basis of tangible interaction ··· 16

    2.3 Theoretical foundation of children psychology ··································· 38

    2.4 Theoretical basis of the design of children's products ························· 54

    2.5 Summary ·· 78

Chapter 3 User Research in the Interaction Design of Children's Products ··· 80

    3.1 Introduction ·· 81

    3.2 Collection and analysis of children's data ··· 83

    3.3 Kidsplay, a research data platform for children's development ············ 123

    3.4 Summary ·· 131

References ·· 133

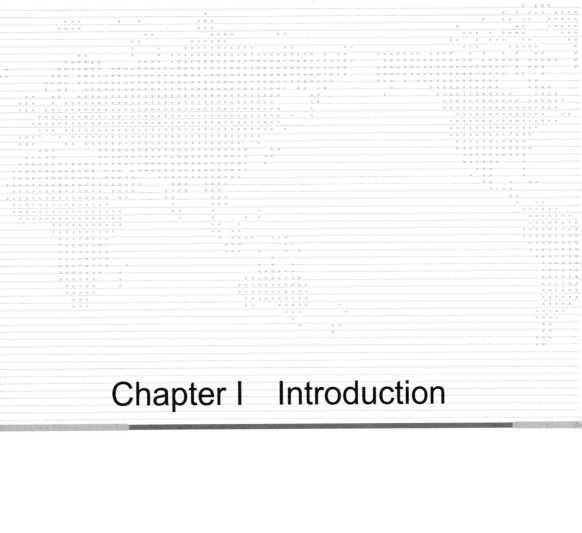

Chapter I　Introduction

1.1　The background of research on the design of children's products

Children in the 21st century have entered the age of artificial intelligence from birth. On the one hand, with the declining age of smart products, children have become the "indigenous people" of the information society. People gradually began to pay attention to the interactive experience of children's products. On the other hand, with the implementation of the second-child policy and the advent of a new round of baby boom, parents in Chinese families have paid more and more attention to the early education of their children. For example, children's education theories such as Montessori, multiple intelligences, Waldorf, Reggio, etc., are widely used in the field of early childhood education in China. New technologies such as artificial intelligence and AR/VR have found application scenarios in the children's industry and have quickly been implemented and promoted. The new demands of personalized education and STEMA have stimulated the emergence of new children's entertainment and education products.

Children are a large and special group. Product development for this group, especially smart toys equipped with smart systems, has received full attention from the market. This book considers such key words as tangible interaction, children's psychological cognition, and the design of children's products, and mainly proposes the following research questions:

How to understand the target users and "design for children".

How to optimize children's interactive cognitive barriers, provide natural interaction methods, and reduce children's learning costs.

How to enhance children's interest in play and learning through design plans, and achieve edutainment effect.

These questions are the main research direction of the design of children's products based on tangible interaction, as well as the overall thinking based on the methodology of physical product interaction design. The problems faced are both opportunities and challenges.

1. New technologies have brought a new ecological chain to the children's industry.

The first element of children's products is to arouse children's resonance. Novel and cool AI interactive technology, lively and interesting educational content, innovative and open thinking, etc., have brought new opportunities and industrial ecological chains to the children's industry. For example, the emergence of image recognition and interactive technology has given traditional tangrams, Rubik's Cube and other toys and teaching aids new ways of playing; the intelligent voice-controlled companion doll with voice recognition, such as the famous "Hello Barbie" can interact and talk with children, which breaks the traditional doll model; the application of VR/AR technology in the field of children's entertainment and education enables children to experience immersive game content. These smart products using new technologies arouse children's interest and curiosity, forming a complete business chain model.

2. The new demands of education require us to come up with targeted solutions and innovative products.

Personalized and customized education is a new trend in the rapid development of the information society. The American New Media Alliance-Horizon Project-publishes the "Horizon Report" every year. The 2015 report pointed out six major trends in the application of technology in the field of basic education. Among them, the rise of STEAM education represents the key development direction of future education. For example, Alt school in the United States promotes a student-centered education model, realizes personalized teaching evaluation and home-school interaction through the school's Internet platform, and implements the education model of "teaching in accordance with the children's aptitude". The rise of the innovative education industry illustrates the attitude of people to embrace science and technology. Many educators and parents hope to introduce these fields into the education model of the children's industry as soon as possible. Well-known open source hardware companies at home and abroad have all launched businesses in the children's education industry.

The scope of the children's industry is broad, and its products not only include daily necessities such as clothing, food, housing and transportation. Entertainment and education products and toys are the most representative category of children's products. Toys, as a kind of simulant and symbol, were not originally made by people for the purpose of games and entertainment. At the beginning of the 20th century, construction games represented by Meccano appeared. The inventors of toys began to declare that toys should "show children the future profession", such as encouraging boys to become workers and builders. At the same time, new material technologies were also widely used in toy production. In 1934, Lego building blocks emerged, which was a kind of hole-column connecting building blocks designed for children. The widespread use of plastic has changed the previous situation of building blocks made of wood. Because of the bright colors and close bite, Lego building blocks have been popular among children. After the Second World War, the sales of toys increased substantially. New materials and new technologies were more used in the toy industry. Toys became cheaper and there were more categories. In 1977, the Star Wars series of toys became popular. Starting from this series of Lucas films, toy manufacturers found that it was a more profitable practice to obtain authorization from movie characters and release new dolls. Pong video games became popular in 1980. In 1984, the cartoon and dolls of *He-Man and the Masters of the Universe* became popular. This was a cartoon made by Mattel to promote the He-Man dolls. The cartoon actually became an advertisement. Since then, a series of cartoons called "program-style commercials" appeared. In 2010, the iPad was launched, and the iPad and smart devices began to become children's favorite game media. In the same year, the maker movement began to become popular in the United States. More and more American educators and families believed that hands-on learning methods and children's interest in technological engineering were very important.

  Tangible interaction is the forefront of the new generation of human-computer interaction. Interactive children's products focus on designing entertainment and education products that conform to children's cognitive models and have high usability. Increasing children's play fun and learning interest is the main problem to be solved. The research significance of this book is mainly divided into the following two points:

① **Theoretical significance:** As an emerging research field, child-oriented physical interactive products need the support of new design theories. Extend the study of tangible interaction behaviors to the field of the design of children's products, fully understand the limited cognitive level of children, and guide children to complete the interaction without barriers, so as to minimize the load of children's learning and cognition, and enhance the entertainment and educational effects of games. Tangible interaction is the foundation of future children's human-computer interaction technology. The integration of physical interface, voice recognition, gesture interaction, various sensors and other technologies is no longer restricted by single-channel interaction. This book explores the interaction system model and the principles of interaction design of children's products, and provides designers with guidance when designing children's products, so as to help children interact with products in a more humane way.

② **Practical significance:** Conducting experimental research on cognitive psychology and emotion in children's games, and providing innovative design solutions for a variety of children's products. At present, many children's industry organizations and children's product markets blindly pursue the upsurge of science and technology, and develop products that do not meet the physical and psychological needs of children. Some of the interactive technologies are not suitable for young children and are difficult for children to grasp. Therefore, when designing child-oriented physical interactive products, how to solve the technical obstacles to make the interactive mode simpler, easier to master, and more likely to stimulate children's interest in games is a problem to be solved. The excellent design of children's products does not require much explanation and guidance to attract children to quickly master the interaction methods and functions, and stimulate children's learning interest and instinct of exploration. Through the study of children's cognition, psychology, behavioral observation methods and experimental methods, this book understands comprehensive aspects of children's characteristics from multiple dimensions, and reconstructs children's cognitive models and user's models. Through the practice of innovatively designed children's interactive products, this book further verifies the feasibility of the design model.

1.2 Domestic and foreign cases

1. First of all, there are many cases of the design of children's products research based on tangible interaction in foreign countries.

For example, the Massachusetts Institute of Technology Media Lab has developed a set of smart building blocks called Topobo[①]. Users can freely assemble the blocks

---

① Topobo is a building block toy. It can make toys that move automatically, but it has a kinetic memory function. This means that these little monsters that you put together can remember the law of your 'fiddling' their joint motions, and can continue to move in the way you've 'fiddled' before powering on.

into various shapes at will, and can control how the blocks move. Topobo blocks are mainly composed of static blocks and dynamic blocks. The dynamic bricks are embedded with motion sensors. As long as the dynamic blocks are assembled at the joints of the object, the system will automatically record the set actions, such as rotation, swing, etc. When the user splices the shape of the building blocks, the system will automatically move according to the movement setting. In this way, a robot dog that can walk, an insect that can swing its antennae, and an earthworm that can twist its body are formed.

Augmented Knight's Castle is a mixed reality interactive platform based on physical user interface. The castle is designed as a medieval castle. The props in the scene design include the RFID[①]—equipped castle, landscape plants, soldiers and kings, etc. Each prop can be accurately located and identified. According to the RFID signal identified by radio, when soldiers with built-in sensors pass by the gate of the castle equipped with RFID, the interactive platform will emit the sound of the army horn; when the king enters the city gate, the interactive platform will emit a salute welcoming the king.

Nintendo Labo. In 2018, Japan's Nintendo Games Company announced

---

① Radio Frequency Identification, or RFID technology, is also known as radio frequency identification. It is a communication technology that can identify specific targets and read and write related data through radio signals without the need to establish mechanical or optical contact between the recognition system and specific targets.
② Bohn, Jurgen. The Smart Jigsaw Puzzle Assistant: Using RFID Technology for Building Augmented Real-World Games. Workshop on Gaming Applications in Pervasive Computing Environments at Pervasive 2004

"Nintendo Labo", which is a new interactive experience for Switch consoles. Children can build a variety of Switch accessories through the current handmade accessories, and insert the Switch device into the built card trays of different shapes. In this way, Switch is given a new mode and new form, and children can experience different experiences such as piano, fishing rod, racing, robot and so on through the brand-new game console. Children can assemble by themselves according to the installation prompts in the game, and design the appearance by themselves.

2. Most domestic researches in this area are concentrated in the field of computer science and technology.

For example, Zhejiang University demonstrated the "Run, Little Chick" in the 2009 National Undergraduate Innovation Experimental Program. The product integrates hardware information technology and somatosensory interaction design, and controls the speed of the chick by means of body interaction through wearable acceleration sensors. When a child imitates the walking posture of a chick-waving his arm to control the movement of the chick, the faster the children swing their arms, the faster the chick runs. Somatosensory interaction exercises children's physical coordination ability, and allows multiple children to interact and compete at the same time, which enhances the fun of the game, and also strengthens the interaction and communication between children.

Grape Science and Technology, a domestic company focusing on children's technologies, has launched "Tao Tao Go Right". The ancient tangram meets the iPad, and the wonderful adventure story allows children to conquer levels in the adventure;

rich tangram graphics allow children to enjoy 700 tangram graphics in 8 categories. Through a high-tech image recognition system, physical operations are presented virtually, children complete tangram puzzles offline, and take an adventure with Taotao, their online wizard partner. In the process of solving each difficulty, the quality of perseverance is deeply rooted in the children's heart, and concentration, creativity and spatial intelligence will also be greatly improved.

1.3 Research content and research methods

1.3.1 Research content

1. Systematic literature research on the theory of tangible interaction interface.

The tangible interaction interface is the result of the tangible interaction design. This book studies the interaction model between children and the tangible interaction system, and on this basis, designs an interaction method suitable for the target situation, so as to meet the needs of children for using tangible interaction systems. The realization of the tangible interaction interface involves many disciplines, but there is a lack of comprehensive research based on the art of design. Therefore, this book will comprehensively sort out and research related theories and their applications under the framework of tangible interaction.

2. Analyzing children's physical and psychological cognitive characteristics, and studying human-computer interaction methods suitable for children.

This book introduces professional theories and research methods such as child psychology, pedagogy and cognitive disciplines to understand children's physical development, cognitive preferences, play behaviors, emotional characteristics, etc. Through experimental observation method, in-depth interview method, questionnaire analysis method, etc., this book collects children's physiological, psychological, cognitive and game characteristics data, and uses the collected data to build a child development research data platform, so as to analyze the feasibility of interactive behavior in children's natural human-computer interaction.

3. Proposing interactive system models and interaction design principles applicable to children's products.

Based on the previous theoretical and basic research on related subjects and child user research, this book builds an interactive system model based on children's products. This model refines and describes the interaction design methods and processes of the entire children's product system from the perspective of target users and designers. The interactive system model provides a theoretical framework for the research of the design of children's products.

4. Designing and practicing children's products and conducting usability evaluations.

According to the interactive system model and design principles of children's products in the previous article, this book guides students to design and practice interactive products for children. At the same time, the book analyzes the hardware and software design of smart products one by one, including hardware modeling, interactive behavior, interactive scene description, and software interface design. Finally, this book conducts usability testing and evaluation of the design practice plan, and provides iterative design support for the interactive system model and product design plan through feedback data.

1.3.2 Research methods

The research methods of this topic include: ① Literature analysis. Extensively consulting domestic and international design research results in the field of children's interactive products and artificial intelligence technology, raising research questions, and seeking theoretical basis. ② Empirical analysis method. In-depth interviews, field surveys and case analysis methods are used to collect and analyze data on children's behavioral characteristics and needs, interaction processes and other issues, and derive a system model of interactive products. ③ Experimental method. The researchers invited the subjects to participate in the experiment in the pilot school, collected cognitive data such as children's interaction behaviors and facial expressions, and established a child development research database. ④ Statistical analysis. Using one-way analysis of variance, T-test, factor analysis, etc. to refine experimental data and analyze the characteristics of children's cognition. ⑤ Practical research. Applying the theoretical research results to the design of children's interactive products, and

Chapter I Introduction

testing its feasibility in actual teaching, as shown in Table 1.1.

Table 1.1 Research methods

Stage	Research method	Explanation of the research method
Raising questions	Literature research	Extensively consulting domestic and foreign research cases related to artificial intelligence, child psychology, human-computer interaction, product design, etc., to grasp the research status and problems.
	Interdisciplinary theoretical analysis	Finding relevant theories from disciplines such as design, information technology and child psychology, analyzing the causes of this problem from different disciplines, and conducting theoretical research and innovation.
Research questions	Experimental method	Observing the interactive behavior of children in the experiment, and collecting data such as interactive behaviors and cognitive preferences.
	Empirical research	Through the extraction of children's interactive behavior rules, as well as the analysis of the game process and the content of the game, this book seeks to find a breakthrough in the model of children's product tangible interaction system.
	Statistical analysis	Using one-way analysis of variance, T-test, factor analysis and other methods to summarize experimental data and transform them into research results, and build a database of children's play behaviors and expressions.
Solving the questions	Practical research	Applying theoretical research results to practice, designing interactive products for children and testing their feasibility.

Chapter 2 Theory of the Interaction Design of Children's Products

Chapter 2 Theory of the Interaction Design of Children's Products

2.1 Introduction

This chapter focuses on the relevant basic theories such as the theory of tangible interaction, the theory of children's development (taking pre-school children as an example), and the theory of the development of children's products, and it also sorts out and analyzes the related theoretical development on the basis of literature research. In this way, it lays a solid theoretical foundation for the subsequent design practice of children's products based on tangible interaction. This chapter first starts from the principle and development history of tangible interaction, and summarizes the development context of tangible interaction interfaces such as ubiquitous computing, graspable user interfaces, physical bits and materialized user interfaces. Subsequently, based on the analysis of the tangible interaction interface of natural behavior, the author puts forward the concepts of low-dimensional tangible interaction and multi-dimensional tangible interaction, and further elaborates the principle of tangible interaction interface. Then, based on the characteristics of the tangible interaction interface and the analysis of related design cases, the author summarizes the four major design elements of the tangible interaction design of children's products. Then, based on the analysis of children's physical development, psychological development, cognitive development, game development and other characteristics, the author summarizes the user's physical, psychological and game behavior characteristics of 4-6 years old pre-school children as an example. Finally, by sorting out the basic theories of the design of children's products, including the analysis of the basic characteristics of children's products, the research factors of design obviousness, and the study of cultural inheritance functions, the author further understands the connotation of the design of children's products.

2.2 Theoretical basis of tangible interaction

The essence of interaction is the need for human survival and development, and the necessary ability for human to adapt to nature and production activities. The "interaction" discussed in this book specifically refers to "human computer interaction[①]". When the system transmits information to the user or the user transmits information to the system, the interaction occurs. Verplank (2006), a pioneer of human-computer interaction, emphasizes user-centeredness, and divides the issues that need to be considered in "interaction" into three aspects: feeling, knowing, and doing, that is, easily controlling an object, and completing tasks such as switching and controlling drones. O'Sullivan and Igoe put human beings as part of the real world to interact with computers in their physical computing framework in 2004. The idea of physical computing puts more emphasis on the computer as the dominant party of

---

① Human-computer/human-machine interaction is abbreviated as HCI or HMI, which is a discipline of the interactive relationship between the system and the user. The system can be a variety of machines or computerized systems and software. The human-computer interaction interface usually refers to the part that is visible to the user. The users communicate with the system through the human-computer interaction interface and performs operations. From the play buttons on the radio to the dashboard on an airplane or the control room of a power plant. The design of the human-computer interaction interface should include the user's understanding of the system (that is, the mental model), which is for the usability or user-friendliness of the system.

Chapter 2 Theory of the Interaction Design of Children's Products

interaction, to perceive, recognize and change the external world, with the characteristics of free exploration and non-linear narration. For example, a movie traditionally has only one ending, and the information structure and narrative method of new media is rarely a straight line. The audience can freely choose to watch the development of different story lines. Moreover, due to the use of 360-degree full-angle and panoramic deep shooting, the viewing angle of the audience can also be changed as desired. The audience can enter the movie, and can also control the flow and direction of the narration just like playing a game. Winograd summarized three modes of human interaction with the outside world in 2006, which are equally applicable in the virtual world: manipulation, locomotion and conversation; Donald Norman proposed a future trend in 2007—the combination of man and machine. This model pursues barrier-free communication between humans and machines, and ultimately achieves a high degree of tacit understanding between humans and machines.

2.2.1 Principles of tangible interaction

Tangible interaction is a process of digital content exchange that takes physical entities as carriers, presents digital information on physical entities and completes them through interactive operations. Professor Hiroshi Lshii of the Massachusetts Institute of Technology defines tangible interaction as a process of interaction between a concrete entity and abstract information, which mainly includes two parts: container and token, which represent the interface of digital information processing and the corresponding hardware device. The main form of tangible interaction technology is to embed sensors or micro-processing chips into physical devices so that physical entities can "perceive" environmental information and perform data processing. At the same time, software and hardware attributes are designed to conform to the interactive data. Researchers believe that the design of child-oriented physical interactive products should focus on the way children interact with digital information and tangible interfaces, and design intuitive and simple interactive behaviors as much as possible to meet the intuitiveness and ease of use of children's interactive experience. For example, Professor Nicholas Negroponte of the Massachusetts Institute of Technology in the United States proposed in his book *Digital Survival*: "Calculation is no longer only related to computers, but will determine our life experience". Designers embed various microprocessors in children's products or other smart devices (watches, glasses, cars and home appliances), starting from children's cognitive behavior and creating a new experience and lifestyle for users.

1. The proposing of tangible user interface (TUI)

The tangible user interface is an information carrier in which the user completes the process of communicating the content of the digital system directly connected to the physical entity through the overall interactive operation of the physical entity object, and responds instantly through feedback actions. The tangible user interface includes not only the carrier of the information interface, but also the overall interactive communication and information interaction process. The original goal of the tangible interaction interface is to apply natural interaction behaviors to the design of smart products as much as possible, reduce the user's cognitive burden on computer technology, and enhance the fluency of use. The tangible user interface has greatly improved the use efficiency of computers, which is the basic paradigm of modern computers. At present, computer systems in the field of human-computer interaction and interactive modes in the form of physical products have become increasingly mature in people's daily lives. In the next part, the author will sort out the development context of the tangible user interface, and further analyze the design of tangible interaction in detail.

① **Ubiquitous Computing** is the source of the original idea of the tangible user interface. Ubiquitous computing is a breakthrough from tradition. It expects that one day computing and communication technologies will be integrated into the structure of the world. For example, ubiquitous computing can be embedded in the fabric we wear, the structure of a building, or anything we carry or wear. At present, the application and technical research of ubiquitous computing is widely used in various industries, and users are surrounded by ubiquitous technologies. One vision of ubiquitous computing is Ambient Intelligence. The "Smart Dust" developed by Hoffman (2003) of the University of California at Berkeley is such a project. It has been able to produce miniature single chips with a size of 3mm and containing all electronic devices in one. This type of micro equipment has achieved higher robustness and functionality, and was used in wireless sensor nodes and industrial applications. In the MIT laboratory, the responsive environment research team focuses on exploring the future environment from a functional perspective. Their ubiquitous sensor interface consists of a large number of sensor arrays scattered throughout the physical space of the media laboratory. This allows the media laboratory to establish a real-time connection with a virtual experimental space in a virtual world (Second Life). People's representatives in Second Life can see real-time images of real-life in the media laboratory, and can communicate beyond the boundary between the two realities; another example of wireless sensor network is the Siftables project developed by the MIT laboratory. A wireless sensor network composed of small building blocks can obtain energy from external energy sources through electromagnetic induction. When two blocks are swiped at very close distances, the device gets enough energy to transmit its unique ID and sensor readings. The application of ubiquitous computing has brought about the output of a large amount of digital content and the information carrier has become diversified, decentralized, and intelligent, making it smaller, lighter, and more portable in terms of physical volume and form. In the future of human daily life, public space facilities, residential communities, school campuses and other environments, people can make full use of this technology to design new interactive systems.

② **Graspable User Interfaces** were the predecessor of the tangible user interface, and the graspable user interface is narrower than the development of the concept of ubiquitous computing. The graspable user interface allows users to directly manipulate electronic or virtualized objects on physical artifacts. These physical artifacts are a new input device, and the interface mixes virtual and physical forms. The user interface can be grabbed to focus on the input operation of both hands, and it is expected to provide a gapless interface interaction between the physical and virtual worlds.

The graspable user interface provides input and output of the mixed space at the same time. Generally speaking, the graspable user interface tries to include two complex interactive features of space and time. The interactive features of the interface have the following advantages: A. Encouraging two-handed manipulation and interaction. B. Specialized context-aware input devices. C. Allowing users to have more parallel input specifications, thereby improving the computer's expressive and communication skills. D. Using the user's proficient daily behavior as the manipulation method of physical objects. E. Reflecting the traditional internal representation of the computer. F. Using physical artifacts to promote more intuitive and better control and interaction of interface elements. G. Utilizing the user's keen spatial reasoning ability. H. Spatial diversity, each operation behavior is mapped to an independent operation device. I. Collaboration, providing multiple devices to be controlled by multiple users at the same time. For example, Professor George Fitzmurice, Horoshi Ishii of MIT and others proposed the Lego Bricks system in the ACM meeting, which is a conceptual-based interactive system with a graspable interface. It is closely connected with virtual objects through physical "bricks" and realizes synchronous control. The number of "bricks" is not limited, and can be simultaneously manipulated on a huge horizontal display desktop. Based on the behavior characteristics of the user when performing a task, it requires the user to use the finger to manipulate the object flexibly. The tangible user interface draws on the idea of this system. Therefore, the graspable user interface is the predecessor of the development of the tangible user interface, and it is an important way for the graphical user interface to extend to the tangible user interface.

③ **Tangible Bits.** In 1997, the MIT Media Laboratory in the United States first proposed the concept of "Tangible Bits" at the 1997 CHI conference. The interface information of the "Tangible Bits" presents the interactive behavior of the media foreground through visual language, and can allow users to perceive the bit information of the surrounding environment by perceiving the light, sound, airflow and water flow in the background environment. The metaDESK and transBOARD system prototypes explore the front stage where users manipulate bit information through physical entities; on the other hand, the prototype of the ambientROOM system focuses on allowing humans to perceive the background of the environmental media, as shown in Figure 2.1. The metaDESK system design attempts to integrate the graphical user interface into the real world, and physically embodies many popular metaphorical devices (windows, icons, buttons, etc.). At the same time, researchers try to express the elements of the graphical user interface through the language of tangible interaction. For example, "LENS" is an arm-type flat panel display that allows tactile interaction in the form of combining three-dimensional digital window information with physical objects, as shown in Figure 2.2.

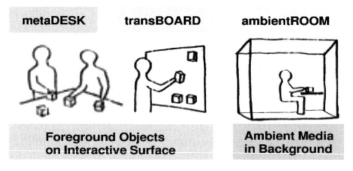

Figure 2.1　Three prototypes of tangible bits

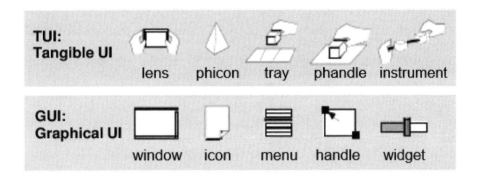

Figure 2.2   The physical use of GUI elements in TUI

The concept of "Tangible Bits" brought the elements of the graphical interface into the real physical world, and developed a prototype of the tangible interaction interface that is different from the graphical user interface, but it did not get rid of the shackles of the traditional graphical user interface. This media theory laid the foundation for the evolution of tangible interaction interface design and created a good academic space for future research on tangible user interfaces. A few years later, Professor Hiroshi Ishii and others created a conceptual model based on tangible interaction and formally extended the concept of tangible user interface (TUI) to the academic field.

④ **Material User Interface, MUI.** The user interface itself is transformed into smart materials that can be changed in real time according to the perception of environmental information. All digital information is completely expressed in material form and directly interacts with human. Based on the development of human-computer interaction technology, the homogeneous use of hardware products is inconsistent with the personalized requirements of users. Users need to use products through "cognitive ability" (that is, the ability to learn, read and understand). This makes the user experience of various products lack of comfort and fun, the satisfaction of the "materialized" accessibility experience is too low, and the interaction methods are monotonous and similar. Therefore, many HCI experts in the field of human-computer interaction are working on exploring new materials for physical interactive products in combination with engineering, art, materials and other disciplines. They have done in-depth researchs and experiments from new deformable and adaptively changing assembly materials to digital information mapping methods.

The Touch Media Group of the Massachusetts Institute of Technology Media Lab put forward the concept of "atomic interface" in 2012, which can solve the problem of the limitation of changing the shape of physical objects in real time, reflecting the "interactive horizon of the future". In 2013, the touch project team released a deformed surface—inFORM, which broke through the traditional interactive method of computer graphical interface and brought users an interesting way of interacting with digital information. Users use screen gestures to control and move real-world objects. The principle of interactive technology is to wave a hand or move an object under the camera sensor. The small movable strips in the hardware will record these actions. At the same time, in the inFORM desktop network system formed, the interactive control can be realized based on the data transmitted by the hardware.

"Materials and touching are the most basic human sensory needs" Hiroshi Ishii mentioned that when humans are in a world gradually surrounded by flat-panel displays, physical touch interaction is trying to prevent human's future from being completely obscured by glass screens. The new type of tangible user interface relies on physical entities, emphasizes user experience-centered design methods, and uses information and communication technology, intelligent interaction technology, new material technology, etc., to make physical interface physical products more intelligent and humane.

2. Nature-based tangible interaction

The interaction between human and the natural world and the interaction between human and machines are treated differently. But today, when electronic consumer products have penetrated into people's daily lives, many tangible interactions between human and machines have become part of human natural behavior. For example, today's pre-school children have been exposed to electronic products very early, and the tangible interaction with these products has even become a part of children's exploration and cognition of the world. The author believes that the tangible interaction with man-made objects is being integrated into human's natural behavior and habits, and is gradually becoming a part of human behavior and habits. Tactile interaction has become the most "natural" pointing device in many occasions, more accurate than eye tracking, and more direct than mouse clicks.

① **Low-dimensional tangible interaction loads visual, auditory, tactile, taste and other feature codes with recognition signals on physical objects to realize the interaction between users and digital information.** Most of the existing tangible interaction interfaces mainly use graphic recognition technology, virtual reality technology, remote transmission technology, etc., to directly identify the information of the physical entity object, and intuitively link the physical representation of the entity with the corresponding digital information content. Most of the low-dimensional tangible interaction forms are relatively simple, with tactile interaction as the main medium of information transmission. For example, the early touch panel Touch Engine allows users to feel the real pressing feeling of vibration when they touch the control panel. The strength of the vibration also varies with the state of the system. Sony introduced the "Smile Shutter" camera. The lens can be used to detect the facial expression information of the person. When it recognizes the smiley expression of the person, it will automatically press the shutter.

② **Multi-dimensional tangible interaction increases the multi-channel way of information input and output, which enhances the multiple ways of perceiving and identifying information.** Human can directly use physical entities as multi-sensory sensors to interact with digital information, allowing intelligent and humanized intelligent entity objects to interact with users in collaboration. In March 2016, Sony launched a new R&D project "Future Lab" and launched a mysterious project: wearable device-smart collar. It allows users to listen to music and receive information freely, while reducing the trouble of putting headphones in their ears. After wearing this money ring, you only need to give voice commands, and it will be able to take pictures, play music and other functional services. And, with the help of motion sensors and GPS functions, it can realize intelligent selection of scenarios. For example, when people are going out from home, it will automatically remind the weather outside, and when they pass the vegetable market after work, it can also remind people that it is time to buy groceries and play a set list of grocery shopping for you; through the interactive projector, users can intuitively interact with the virtual scene through gestures, making the display of objects more interactive. The projector uses a unique image recognition technology and depth sensor. Through machine learning, it can recognize and track people's various gestures (including rotation, tilt angle, movement, etc.), and make corresponding responses. The projector can recognize the 3D space, and interact with the screen projection and speakers, so that people can see the little figures in the fairy tale books move. It can also perform clear projection of objects on the desktop, and can immediately calculate some data of the projected objects. For example, if you put a teacup and a book on the desktop, it can immediately calculate their size and project them on the desktop to display them in a way that people do not notice.

2.2.2 Characteristics of tangible interaction

Tangible interaction is a new type of interaction method that provides interaction design theories and methods that are different from graphical user interfaces. Its design focuses on the physical entity or the physical environment, where users communicate and interact with information through material interfaces. With the development of information technology, the way of tangible interaction design has become more diversified and natural, encouraging humans to further think about how to integrate tangible interaction into daily use scenarios. For example, in the fields of children's education and entertainment, health and medical care, household appliances, transportation and travel, the application cases of physical interactive product design have been widely applied in various fields. At present, the conceptual framework of tangible interaction design is gradually maturing, which helps academic researchers and designers to classify the theoretical models of tangible interaction design and conduct personalized design methods for different types of products. The tangible user interface mainly has the following four characteristics:

1. The physical objects are mapped to the corresponding digital information and used as its manifestation.

The main feature of the tangible interaction interface is the coupling of the entity with the underlying digital information and calculation model. For example, the above model example illustrates the series of relevance, including the use of graphic data to construct entities, simulation calculation control (operation) tools and auxiliary material properties. The physical entity establishes a mapping and matching relationship with the corresponding digital information and the user's perception ability, and interacts through multiple channels such as touch, vision and hearing, so that the user and the information space are seamlessly connected to form a smooth user experience. For example, as early as 2003, Stanford's iStuff tried to replace the interface components in the GUI with real objects, such as pens, sliders, buttons, toy dogs, stylus, microphones, etc. Physical interactive devices were connected via a wireless network, allowing multiple physical devices to work together in the digital toolbar. With the development of 3D and 4D printing① technology, on the basis of 3D

---

① 4D printing, to be precise, is a material that can deform automatically. It only needs specific conditions (such as temperature, humidity, etc.) and does not need to be connected to any complicated electromechanical equipment, and it can automatically fold into the corresponding shape according to the product design. The most critical aspect of 4D printing is "smart materials". 4D printing was developed by MIT in cooperation with Stratasys Education Research and Development Department. It is a revolutionary new technology that allows materials to be quickly formed without a printer.

printing, the MIT Materials Laboratory uses computer 4D printed deformable materials to form a tangible user interface; the assembly robot Roombot developed by the EPFL Bionic Robot Laboratory in Switzerland can move freely and assemble automatically, and can freely change its shape like a building block. By collecting sensor input data from the surrounding environment, the bionic robot can understand the user's current needs, and then transform according to the user's needs. The digital information is combined with the physical interface to realize the synchronous change of the functions of the digital content form and the physical form, and a mapping is established between the bit information and the physical interface, so that the user can directly interact with the physical object.

2. In order to realize the interactive function, the electronic mechanical structure is embedded in the physical entity.

The presentation objects of the tangible user interface are usually seen as a way to control the tangible interaction. The physical movement and rotation of these workpieces, their insertion or connection with each other, and other operations are the main control methods of the tangible interface. The physical user interface integrates the input and output of information into the device, especially the output of part of the information and the input of all information into an operable physical entity. Using physical carriers with digital characteristics, users can intuitively interact with virtual digital information, which improves the experience of interaction. The physical interactive building blocks developed by MIT, Mediablocks, are composed of many small blocks that can store and transmit digital information. The "Mediablock slot" has built-in miniature whiteboards, cameras, printers and digital projection equipments. When a block is inserted into the slot, the information associated with the block is transmitted to the shared network. In addition, the product also contains two manipulators for combining and editing digital information in various Mediablocks, and the computer equipment acts as a gateway between blocks and slots. The whiteboard and camera record the data in Mediablocks in the device, and the printer and projector display the information stored in the building blocks in the browser.

3. Physical entities have the function of "perceiving" digital information.

The physical interface relies on the balance between physical and digital representations. The embodied physical elements play a central role in defining the control and performance of the tangible interaction interface, especially the digital representation of graphics and audio. Usually, most of the dynamic information is processed by the underlying computing system. Generally speaking, interactive products need to directly interact with the physical world, and the interaction of digital information needs to be realized through the carrier of software and hardware. Therefore, in order to have a better user experience of tangible interaction, people use their own sensory organs to perceive the interaction, and combine the ways of human perception through the form of "material" tangible interaction interface. The graduate student of Harbin Institute of Technology designed a music player Tmusic based on a physical user interface. It uses the music emotion type corresponding to the tangible interaction emotion type, and uses various character expression models based on character expressions to correspond to different key music types. For example, when the user selects a sad expression model, the system will play a song with sadness as the main melody; when the user selects a happy expression model, the system will play a song with a happy melody. The designed interaction mode unifies the entity attributes and digital information, allowing users to easily and intuitively grasp the use of the Tmusic entity player; in addition, using the character model as an interactive carrier makes it easier for users to recognize and generate emotional resonance.

4. The characteristics of a physical entity can reflect the characteristics of the digital information it represents.

In fact, the tangible user interface is generally constructed by the physical artifact system. The readability of a physical entity is determined by the physical state of people and computers, and its physical structure is closely coupled with the state of digital information presented by the system. Since the concept of tangible interaction interface was put forward in 1997, the Touch Media Lab of MIT has further expanded and proposed the Material User Interface. That is, the interactive carrier of the physical world can change in real time (including appearance and shape changes) according to the state of digital information. The concept of material user interface extended the research scope of tangible user interface. At present, this kind of research is a new trend in the development of tangible user interface. No matter how complex the characteristics of the material and how flexible the structure is, it can be displayed, reflected, fed back and expressed based on digital information. For example, binary synergistic nano interface materials [①], bionic intelligent interface materials, biomaterials, etc., abandon the tradition and synthesize brand-new materials, and carry out some special processing on the surface of the materials, so that the materials have special functions. It will take a long time for the material interface material technology to mature and enter the market, but we need to try first in terms of the exploration of interaction design technology. The MIT Media Lab has successively launched Radical Atoms, FocalSpace, MirrorFugue, Tangible CityScape, Transform and other projects. They have done in-depth research and experimental development from new deformable and self-adaptive assembly materials, to new energy supply methods, and digital information projection mapping expression methods. In this way, they carried out revolutionary cutting-edge subversive innovations in the interaction of physical user interfaces. Returning to the material world itself has become a new direction in the study of human-computer interaction interfaces.

This section will analyze the characteristics of the tangible interaction interface one by one, and we believe that the tangible user interface can also be realized through

---

① "Dual synergistic nano interface materials" is a new concept. It is different from traditional single-phase materials, but it rather builds a structure of binary synergistic nano-interface on the macroscopic surface of the material. The design idea of this interface material is that people do not necessarily seek to synthesize brand-new bulk materials. When a special surface processing is adopted, two interlaced and mixed two-dimensional surface phase regions with different properties can be formed at the mesoscopic scale; the area of each phase region and the "interface" constructed by the two phases are all nanometers in size.

② Zhang Handan. Research on music emotion visualization system and user experience based on physical user interface [Master's thesis]. Heilongjiang: Department of Mechanical and Electrical Engineering, Harbin Institute of Technology, 2016.

more input modes. In fact, as long as you follow the characteristics of these tangible interaction interfaces analyzed, people can create brand-new interfaces for some existing interaction modes. These interaction modes include: touch and touch, voice control, air gestures and body interaction, etc., thus effectively using human life experience in the real world. As far as the tangible interaction interface is concerned, the designed physical user interface is only an external presentation form. What is more needed is to use the potential of modern interactive technology to better adapt the design of interactive behavior to the user's instinct, adapt to specific environments and tasks, and meet user requirements, which is the essence of the concept of tangible user interface.

2.2.3 Design elements of tangible interaction

In view of the previous research content—the concept of tangible interaction, the development and reform of tangible interaction, the characteristics of tangible interaction and the research and analysis of related tangible interaction cases, this section summarizes the four major elements of tangible interaction interface design.

1. A tangible interface that directly interacts with digital information.

The core technology of the tangible interaction interface is to use computational physics technology to enable users to interact with computer components and sensors through physical entities, that is, to communicate and interact with digital information content through the operation of daily tangible interaction interfaces. For example,

"Urp Urban Planning Simulation System" is a famous case of tangible interaction interface, which studies the actual shadows and reflections formed by wind effects between large-scale buildings. It can simulate the light exposure of day, night and different seasons through the computer, allowing the system to automatically judge and calculate the influence of the shadow of the building on the surrounding environment and the air flow. In addition to workbenches with tangible interaction interfaces, tangible user interface with larger sizes and interactive spaces such as interactive walls are also widely used in public spaces. In some new media interactive art installations, you can see many such works of art, such as "Hole in Space" "ClearBoard" "Rasa" "Senseboard system", etc. The "transBOARD" developed by Ishii and Ullmer in 1997 combined a super card with a magnetic back and an embedded barcode with an electronic whiteboard. The content of the electronic whiteboard is scanned and copied to the super card, stored in metaDESK, and used as the same super card for application.

2. Using reasonable perceptual interaction behaviors.

In addition to emphasizing the direct interaction of digital information, the tangible interaction interface also needs to adopt a reasonable perception and cognitive ability as the user's interaction method, avoiding the WIMP (window-icon-menu-pointer) interactive mode of traditional graphical user interface, and removing the disadvantages of "computerization" and "homogenization". For example, the tangible interaction interface "Rainbottle" uses the familiar way of opening and closing the bottle stopper to map the action of turning on the radio switch. The manifestation of "Rainbottle" achieves the cognitive effects of synesthesia on the concrete perception of music. The interactive form of experience from life provides a zero-barrier interactive experience for users' cognitive understanding. The RIMA designed by German designer Matthias Pinkert is an LED table lamp. The product breaks through the traditional button switch design, using a ring controller to control the switch and light of the lamp. When the user slides the ring controller, the LED between the ring controllers can be lit and the color of the light can be controlled. This interactive method does not require users to learn and is conducive to understanding. The user can control the RIMA lamps by sliding the ring controller only by relying on natural perception.

3. Emotional-driven interactive expression of products.

Emotion is the biggest characteristic of human beings, and emotional design has a very important position in interactive product design. The American interaction design expert and design psychologist Donald Norman proposed three levels of emotional design in the book *Emotional Design*, namely the instinct level, the behavior level and the reflection level. Norman emphasized that the instinct layer and the behavior layer respectively represent the attributes of the product and the user's behavior, while the reflection layer covers information, culture and product efficacy. The most essential attribute of a physical interactive product is to meet the user's emotional experience needs, and it is the biggest driving force for the target user to interact with the product, rather than a cold functional product. Pay attention to emotionally driven product expression that can resonate emotionally with users, perhaps a memory, a story or an emotional experience. The "things" mentioned here are no longer metaphysical products, but lively and vivid emotional stories, allowing people to have an immersive experience of life style, rather than the alienation and mechanical feeling that complex technology brings to people. Researchers are exploring the new generation of human-computer interaction from the technical level, and are especially trying to design the operation behavior of the product from the way of emotional cognition. It is particularly important to map behavioral cognition and emotion in the most familiar way, so that they can maintain their initial simple needs and enthusiasm, and bring people a pleasant user experience. In 2005, well-known designer Wensveen designed an affective alarm clock, which is positioned as an "alarm clock that recognizes human emotions". Products include 12 sliders, two displays, etc. The display at the bottom of the alarm clock displays the time, and the display at the center of the circle displays the alarm time preset by the alarm. Users can slide multiple sliders to express their own emotions. According to the change of operation behavior, the different position of the slider will form different graphics, thus forming a graphic that remembers and tracks the user's operation. The designer applies the learning and conditioning theory in psychology to judge the user's emotional changes based on the graphical trajectory of the user's operation. On the other hand, the "Cloud of Cloud" lamp designed by the graduate students from the China Academy of Art uses cloud shapes as elements to give the light a more natural and true feeling. By using childhood paper airplanes to glide over the lower end of the clouds, users can light up the "Cloud of Clouds" lamp. "Cloud of Cloud" associates the chandelier with the shape of the cloud from the emotional design, giving the product a vivid form, which has both functional and decorative features. At the same time, in terms of interactive forms, the switch mode of the paper airplane brings users into childhood memories, and the lively and interesting forms evoke different emotional memories of users to varying degrees.

4. Interesting user experience design.

User Experience (UX) usually refers to the subjective feelings and feedback generated by users when using a certain product, system or service. User experience includes emotions, beliefs, preferences, cognitive impressions, physiological and psychological reactions, behaviors and feedback generated before, during and after use. Products with a good user experience are surely associated with users' cognitive impressions and emotions. For physical interactive products, creating an interesting user experience is not simply fun, but associating the experience design with the attributes of the product itself. Compared with more complex physical interactive products, the more functions of the product and the more independent the defined attributes, the more difficult it is to create an interesting user experience. Therefore, when providing users with a pleasant experience, it is necessary to consider whether each feature, function or step of the product will increase the load of the user experience. For example, in 2014, a Japanese designer invented the "Escaping Wallet". This kind of smart wallet has built-in sensors and is associated with the mobile app. It has a money-saving mode and a consumption mode, which connects the physical world and the digital world. When in the "money saving mode", if the user reaches out to grab the wallet, it will automatically escape and make a call for help; when in the "consumption mode", the wallet will proactively approach the user and play Amazon's product sales ranking to stimulate user consumption. Although this kind of interesting experience technology is not yet mature, it is in line with user experience expectations, and its development prospects are very optimistic.

Starting from the analysis and induction of the design elements of tangible interaction, this section introduces four related design elements and their related application cases: Interactive interface that directly interacts with digital information; using reasonable perceptual interactive behavior; emotionally-driven product expression; interesting user experience design. We have seen that whether it is to provide intuitive interaction methods or to meet human perception and emotional cognition, the design of tangible user interfaces is essentially to return to the essential attributes of "things" and to acquire the emotional needs of users. This is a reflection on the "emotional desertification" of most products, and at the same time we need to pay attention to the thinking of information design in the digital world.

2.3 Theoretical foundation of children psychology

Children psychology is a basic theoretical discipline that systematically expounds the basic laws of the psychological development of children from birth to adulthood. The developmental age group of children psychology is divided into 0-18 years old. Infancy refers to from birth to 1 year old, early childhood refers to from 1 to 3 years old, and pre-school refers to 3 to 6-7 years old, school age refers to entering puberty from 6-7 years old to 12-13 years old, and puberty is 12-13 years old to 18 years old.

This chapter takes pre-school children aged 3-7 as an example. Children of this age are easy to learn knowledge and skills or form psychological characteristics. Good guidance can promote the rapid development of children's psychological level. If this period is missed, children's learning and development will be slower. Therefore, the study of children psychology helps to understand children's development laws, enrich the general theories of design psychology, provide theoretical support for the design of children's products, and promote the development of children's educational theories, thus better providing convenient services for the development of the children industry.

2.3.1 Characteristics of children's physiological development

The physiology of children is different from that of adults. They have the peculiarities of physical development, which is also the basis for understanding child users. For example, pre-school children are quite different from adults in terms of external organs, motor ability, language ability and nervous system. The researcher consulted the children development research report on the physical development phase indicators of pre-school children. For example, the average weight of pre-school children aged 3-7 is 12.3-27.8 kg, and the average height is 94.1-125.4 cm. At this stage, they have learned to sit upright and stand upright by themselves, have a certain degree of athletic ability, and have flexible and coordinated hand movements. Pre-school children have perceptions such as sight, hearing, smell, touch and mathematical spatial concepts such as size, distance and orientation. They mainly rely on perception to collect and recognize information and accumulate cognitive experience.

1. Children's nervous system.

Children's nervous system is more sensitive than adults, and has a stronger stress response to external stimuli, and is easily disturbed by various external factors; on the other hand, children's brain development is in a period of rapid development, and the development of brain thinking is extremely energetic, which is the peak time for intellectual development. However, children with late brain neuro development are more distracted, especially the younger the children, the weaker the control ability. Therefore, the design strategies for the neuro developmental characteristics of pre-school children are summarized as follows: ① Developmental characteristics: The forehead area responsible for tissue analysis is developing rapidly; the left hemibrain is more developed than the right hemibrain. Children of this age have keen observation skills, strong learning abilities, and are easily distracted. ② The design of children's products strategy: According to the analysis of the developmental characteristics of the nervous system of pre-school children, product design for pre-school children needs to exercise children's game skills and strengthen language communication skills; the interactive behavior needs to be consistent with the children's language, movement and perception, so as to effectively improve the children's perception level; the development of the left and right brains is not balanced, which requires designers to consider how to use products to make children's left and right hands develop at the same time, which is conducive to balancing the development of the left and right brains.

Chapter 2 Theory of the Interaction Design of Children's Products

2. Children's external body organs.

Children's body organs have the following characteristics: the tactile organs are sensitive, the body and skin are tender, and vulnerable to injuries. Pre-school children's auditory organs are not well developed, but they are very interested in musical melody and rhythm; their visual organs are immature and their eyes need to avoid direct sunlight and develop good habits; their muscle attraction is still developing, the large muscle movements are more flexible and coordinated (including hand muscles, leg muscles, etc.), and the fine movements of using small objects are becoming more and more proficient. In terms of muscle coordination, sense of balance and agility of small muscle skills, children can already use small tools proficiently. Therefore, the design strategy for the development of the body organs of pre-school children is summarized as follows: ① Characteristics of body organ development: Pre-school children's perception organs develop rapidly, their perception abilities are sharper than adults, and their big movements are mature, but their fine movements need to be perfected. Compared with adults, they still need to improve training of life skills. ② Strategies to pay attention to when designing children's products: The size of the product for pre-school children needs to conform to the physical characteristics of children of that age. For example, the size of the product should be suitable for children to grasp, but avoid objects that are too small to make it inconvenient for children to grasp, the product design should be designed to match the children's physical development, and safety issues should be paid enough attention.

3. Children's motor characteristics and skills.

The characteristics of children's motor skills include: they like jumping, running, crawling, throwing, etc., their motor nerves are active and energetic. As they grow older, they show greater differences. For example, pre-school children spend most of their time in non-stop exercise, especially boys who like to explore the space environment with various big movements; in terms of fine and small movements, they can button clothes, string beads, eat on their own with a spoon, hold a pen to paint, put together building blocks, jigsaw puzzles and conduct other activities. Therefore, the product design strategy for the sports characteristics of pre-school children is summarized as follows: ① Movement characteristics. Great motor skills include running, jumping, crawling, throwing, etc. Fine movements are mainly reflected in games and activities such as drawing, building blocks and jigsaw puzzles, and the development of motor ability is more active. ② Strategies to pay attention to when designing children's products: Designers need to fully consider the active characteristics of children at this age, and use appropriate interactive behaviors in the design of children's products to mobilize children's enthusiasm for interaction with the product; at the same time, interactive gestures need to match the fine abilities of pre-schoolers in order to achieve the expected game effect.

4. Children's language and thinking skills.

Children's language expression has features such as situational and repetitiveness, and the logic and coherence of the language are weak. Especially the language expression ability of pre-school children develops rapidly during this period, which is mainly manifested in the rapid increase of the number of vocabulary, the rich and changeable content of the vocabulary, the wide extension of word types and the continuous increase of short-term memory. Pre-school children have outstanding unconscious memory capacity, concrete memory is better than abstract memory, image thinking is better than logical thinking, and children's memory capacity increases with age; on the other hand, because pre-school children's brain cells are active, they have better divergent thinking ability, and have the ability to perceive information in concrete thinking. When children are about 5 years old, their abstract thinking ability is gradually formed, and they begin to think about things simply by reasoning, and classify and think from different perspectives. Therefore, the design strategies for the language thinking characteristics of pre-school children are summarized as follows: ① This age group is a critical period for language development, and expressive ability improves quickly. ② Memory becomes easy, but the accuracy is not high. Image memory is a characteristic of the period, and it is necessary to repeat the memory to improve the memory effect. ③ Issues that need attention in design: The design of children's products should exercise children's language expression and communication skills as much as possible, and use figurative modeling techniques to enhance memory effects, increase interest, and drive children's emotional memory characteristics.

2.3.2 Characteristics of children's psychological development

The design of children's products should not only meet the physical needs of children, but also take into account the psychological needs of children. Good product design should be able to meet the different psychological needs of children and have a positive psychological impact on them. The main psychological characteristics of children are curiosity, unstable interests, eagerness to try everything, personal experience of everything, strong thirst for knowledge and cognitive interest. Compared with rational and serious adult furniture, children prefer combination furniture with changeable play styles. This set of children's furniture in the picture has cute shapes, and at the same time provides a variety of play functions for children, which fully meets the psychological needs of children at this stage.

1. The developmental characteristics of children's curiosity.

Children will have a strong sense of curiosity about novel environments and new things. They like to explore novel things through their own tactile and auditory senses to satisfy their curiosity. In particular, pre-school children lack life experience and common sense, and most of their cognition of new things comes from asking questions to their parents. They will ask many questions, and at the same time, they will personally touch, play and try new things. Curiosity has positive meaning for the growth of young children. It is a manifestation of children's love of thinking and learning, which is conducive to the healthy development of the body and mind. Pre-schoolers are eager to explore, have rich imagination, and like to slowly explore the origin of things; as they grow older, children develop from asking questions at the beginning to actively touching, observing and dismantling. At the same time, as children improve their knowledge of the growth environment and cognitive abilities, their curiosity will gradually weaken.

2. The characteristics of children's companionship.

As children grow up over time, the objects of children's communication gradually develop from parents and relatives at home to playmates, kindergarten teachers, etc. Pre-schoolers like to play with older children and imitate them. But the older children are reluctant to play with the younger children. The children use their brothers and sisters of the same age as their role models for learning and imitating. At the same time, the circle they like to associate with will gradually change with age. Children's interpersonal communication centers on the school and extends to the neighborhood, because this environment tends to gather many peers of the same age. Communication with peers can meet the needs of children's communication and social responsibility, and at the same time provide a communication channel for children to learn from each other, which is an important source of emotional appeals in children's psychological development. According to related children psychology experiments, it can be found that from the age of 3-4, pre-school children prefer to play with same-sex peers, mainly because same-sex peers have common interests and common topics. The degree of children's attachment to companions increases significantly with the increase in the number of friendships established by companions.

3. The characteristics of children's personality.

There is a Chinese proverb saying that "seeing adulthood from what a child is like at three years old and seeing old years from what a child is like at seven years old".

Pre-school children have formed innate personality characteristics, and at the same time, children's personalities will also change with changes in the environment and age. According to research on children psychology, children's psychological characteristics have been developed since the age of 3, and at the age of 5, children have formed obvious personality characteristics. In terms of self-control ability, the age of 4-5 is a turning point in development. By the age of 6, most children have a certain degree of self-control. Self-emotional experience is also constantly deepening development (happiness, anger, grievance, self-esteem, shame, etc.), among which children aged 4-4.5 can experience emotions such as self-esteem and shame. At the same time, due to the differences in children's family growth environment, hobbies and cognitive abilities, they begin to show greater differences in their personality characteristics. For example, some children are outgoing and self-confident, while others are inferior and withdrawn. When designing products for children, designers need to conduct in-depth research on the physical and psychological characteristics of children, design suitable products for them, and help them grow up healthily and happily.

4. The characteristics of children's imitative ability.

Children have a strong ability to imitate. They imitate everything in the cognitive world, for example, they learn communication and life skills by imitating the language and behavior of adults. Children over 3 years old gradually develop from manipulating objects to consciously imitating simple movements that appear in front of them. For example, they will try to imitate when they see other people drawing. When the children's behavior of manipulating the object enters the function key of the object, the delayed imitation of the behavior appears. With the development of children's perceptual memory, the perceptual image memory stored in the mind will trigger the imitation of the original image due to the stimulation of life scenes. Children's cognition of graphics precedes their understanding of words, and humming a song to the music melody precedes memorizing the lyrics. They haven't had systematic pinyin learning, but are able to learn language expression. They do not recognize characters and understand the meaning of poems, but they can memorize poems backwards. This is the result of children's imitating learning.

2.3.3 Characteristics of children's cognitive ability development

Cognition is an important part of children's development, and the development of cognition is a process of continuous improvement of human's information processing system. The formation of children's cognitive structure and cognitive ability changes regularly with age. The development of children's brain structure and bodily functions, the development of imagination and logical thinking abilities, are of great significance to the development of children's cognitive abilities. Children's products help children improve their logical thinking and cognitive development abilities.

1. The developmental characteristics of children's perception

Perception is the initial stage of children's cognition of things. On the basis of perceptual behavior, children can form a further understanding of new things, thus forming a series of more complex psychological processes and acquiring the ability to learn and understand themselves and the outside world. Children are more likely to perceive the world through the senses such as graphics, colors, materials and sounds, rather than through language and behavior.

① **Children's visual development.** Visual stimulation is the most important source of information that forms human cognition. Taking the visual development of pre-school children as an example, they mainly show the following characteristics: A. The precise development of visual acuity. Children's vision for distinguishing detailed objects and comparing object differences ranges from black and white and blurry at birth, to precise, colorful, and clear. Visual acuity continues to improve with age, and children's visual acuity does not develop completely until 6 years old. Therefore, the pre-school period is an important period for human vision development, and eye diseases such as myopia, amblyopia, and strabismus are easily formed and best treated at this age. B. Color discrimination refers to the ability to distinguish subtle differences in colors. Research findings have shown that the correctness of color recognition and the types of color recognition of pre-school children increase with age. For example, a 3-year-old child can already recognize basic colors (red, yellow, blue, green, etc.), but it is difficult for them to recognize mixed colors. Children aged 4-5 can recognize mixed colors (pink, earthy yellow, dark green, etc.), and children aged 6 can accurately recognize the colors they touch (gray, sky blue, bronze, etc.). Due to the lack of life experience of children, it will be detrimental to the development of children's visual ability if there is no adult to carry out the correct guidance.

Chapter 2 Theory of the Interaction Design of Children's Products

② **Children's hearing development.** The main channel for children to recognize sounds is hearing. They rely on hearing to understand the sound characteristics of things. For example, children can perceive the melody of music through hearing, learn various languages, watch favorite cartoons, and improve their intelligence. The analysis of children's hearing characteristics is as follows: A. Children have a keen sense of hearing. Children's hearing acuity differs due to individual differences. Generally speaking, their hearing sense is extremely keen, and they like to receive interactive information of sound, but the physiological structure of the ear organs will gradually develop and perfect with age. B. The importance of speech and hearing. The talents of pre-school children to learn languages far exceed those of adults. They can distinguish the nuances of language and pronunciation in the middle stage of pre-school, and they can master the national language or other languages in the later stage of the pre-school stage. C. Cultivation of the safety of hearing. Children's auditory nerves are relatively fragile and are easily stimulated by noise to affect the development of auditory nerves. Parents can protect children's hearing development by creating a relaxed environment, and at the same time, they can also improve children's language expression ability through systematic language training. Children's hearing affects the development of language ability, thinking expansion and interpersonal communication skills. Parents should attach great importance to the cultivation of children's hearing ability.

③ **Children's tactile development.** Tactile sensation is the sensation produced when the skin is mechanically stimulated. As an important means for children to recognize the world, tactile sensation plays an important role in the development of childrens cognitive ability. Children have tactile responses to the outside world from birth, such as sucking, grasping, and dragging reflexes among the unconditional reflex behaviors. Their main tactile organs are the mouth and hands, and the hands are the main channels through which they know the outside world through touch. The appearance of coordinated eye-hand movements indicates the beginning of the real tactile exploration of the hands. Children mainly perceive the basic attributes of objects by touching physical objects with their hands, such as the softness of cotton, the hardness of fruit shells, the smoothness of eggs, the roughness of stones, the coldness of ice and snow, and the hotness of campfires. Product design for pre-school children should fully mobilize children's tactile senses, and develop their tactile perception abilities by using material changes, styling bumps and grasping strength, thus improving children's tactile perception ability in the interactive contact with toys.

④ **Children's perceptual development.** Children's perception mainly includes space, time, movement, etc. A. Spatial perception is children's perception of the spatial relationship of things. For example, children's perception of three-dimensional modeling is the result of the combined effects of vision and touch. According to experimental research, children's perception of shape recognition generally develops as follows: 3-year-old children can correctly recognize basic geometric figures such as squares, rectangles and triangles, 4-year-old children can correctly identify semicircles and trapezoids, 5-year-old children can correctly identify various geometric figures such as rhombus, parallelogram, ellipse, etc., and master the names of basic shapes.

Chapter 2　Theory of the Interaction Design of Children's Products

The development trend of children's sense of direction is: children around 3 years old can distinguish up and down, children around 4 years old can distinguish left and right, and children around 5 years old can distinguish the north, south, east and north sides centered on themselves. B. Time perception is an important concept in psychology. The human brain's reflection of the continuity and order of objective phenomena is also called the sense of time. From the age of 3, children perceive the concept of time through the connection of the order of things. 4-year-old children have fully understood the concepts of yesterday, today and tomorrow, but cannot distinguish between temporal and spatial relationships. Children aged 5-6 can recognize the order within a day and within a week, and learn to distinguish temporal and spatial relationships. C. Motor perception is a child's perception of the spatial displacement and speed of objects. Through the perception ability of movement perception, children can clearly distinguish the movement state and movement speed of objects. According to the physiological characteristics of children's hyperactivity and distraction, designers can consider designing interactive toys that meet the children's motor perception ability.

2. The developmental characteristics of children's observation ability

Children's observation ability is mainly based on vision as the center of activity development. Combined with the behavioral characteristics of gestures and pointing, visual perception should be guided by hand movements, and the order of children's observation gradually develops from skipping and disorderly behaviors to orderly observation. The characteristics of children's observation ability are summarized as follows: ① The basic characteristics of their observation ability: Children are easy to choose dynamic, vivid, bright, large and clear objects as observation objects. For example, the cartoon dolls in the children's playground always attract most children to stop because of their bright colors, vivid shapes and large volumes. On the other hand, research has found that objects with obvious characteristics are easy to be observed, while objects without obvious characteristics are easy to be ignored. For example, pre-school children can easily remember the difference in the size of the ball, but cannot remember the difference in colors and patterns. ② Cultivation of observation ability: A. Mobilize children's multiple senses to observe things. For example, children can be brought to nature, to perceive the beauty of nature, to see the beautiful colors of various flowers, plants, fish and insects, to feel the cold of the lake and the warmth of the sun, and to listen to the sweet birdsong, so as to recognize the features of climate changes. B. Guide the children to observe the details in depth. Cultivate children's systematic observation and develop better study habits. For example, observing insect specimens can follow the observation principle of from the outside to the inside, from the whole to the part, and from the obvious features to features less obvious. C. Provide a good environment and observation objects for young children. As the main places for children's life and learning, homes and schools need to provide a wealth of observation materials and guide children to participate in observation with multiple senses.

3. The developmental characteristics of children's attention

Attention mainly includes two basic forms: unintentional attention and intentional attention. The development of children's attention is mainly based on unintentional attention, and intentional attention is gradually formed in the later stage. The characteristics of children's attention development are summarized as follows: ① The unintentional attention of young children is dominated by unpredictable and unrestricted attention. They are mainly attracted by the physical characteristics of everyday sudden objects. ② Young children are at the initial stage of intentional attention development, which refers to the control of willpower to guide the development of attention. Due to high instability, they need to rely on adult guidance to cultivate intentional attention. In the early stages of pre-school children, their attention span is very short, and their attention to something is not long, and are especially easily distracted by the surrounding environment. Parents or teachers can guide them to learn to observe something continuously. Therefore, when designing products, designers need to consider how to help children concentrate and attract the attention of young children. Designers can highlight the features that need attention in children's products and focus on them to cultivate children's concentration.

4. The development characteristics of children's imagination

Children's imagination is the psychological process of children's processing and transformation of old representations to establish new images. The development of imagination is one of the important characteristics of children's cognitive development. They can use imagination to build their own game world, learn and recognize everything, and then promote children's physical and mental health. Children's imagination is mainly divided into: re-creating imagination and creative imagination. ① Re-creating imagination. It refers to the process by which children form an image of a certain creative thing based on certain reference things (graphics, charts, symbols, language, etc.). ② Creative imagination. It refers to the process of children independently constructing new image things according to their own ideas. Studies have found that creative imagination comes from unconscious divergent associations. The images of creative imagination are often very different from conventional prototypes. The scenes of creative imagination are gradually enriched because they are not bound by cognitive experience. Due to the limitation of children's cognitive level, the limitation of imagination performance, and the influence of emotions on imagination, children's imagination has the gift of creating exaggeration. They like to build exaggerated images and content in a way that breaks through the conventions, and they often confuse reality with creativity. Therefore, cartoon characters tend to adopt exaggerated forms and bright colors to strengthen and exaggerate certain features. For example, the popular images of Mickey Mouse and Donald Duck are lively and interesting, allowing children to learn and understand the world in a happy atmosphere, making it a cartoon that is entertaining and suitable for children to watch.

2.4 Theoretical basis of the design of children's products

With the advent of the technological boom and the investment of a large amount of capital, the intelligentization of children's products has become a trend. However, these children's products lack serious thinking about children's groups, and

Chapter 2 Theory of the Interaction Design of Children's Products

unreasonable functions and wrong values are flooded with inferior children's products. The children's product market is a consumer product market full of vitality and business opportunities. The global children's product market is very large, with diverse product types, and a very broad demand for design. Among the internationally renowned children's products, there are brands that specialize in the production of children's toys. For example, the Danish LEGO, which is popular all over the world, has a history of 80 years. Its products include depicting the "Star Wars" series, the "Batman" series, the "Architect Bob" series, the "Toms Train" series, and larger Duplo blocks suitable for infants and young children; there are also brands involved in the children's animation entertainment market. For example, the American Disney Company has slowly developed from an initial film distribution company to an animation film company, and has turned to the world's largest entertainment theme park, and then extended to the market of children's toys. the Disney-themed doll "Mickey" mouse shape has been applied to various functional toys; there are also large-scale comprehensive children's product stores, such as the world's largest toy chain "Toys "R" Us" (Toys "R" Us[①]), which integrates various types of children's brands to provide consumers with a comprehensive, one-stop shopping experience.

Children generally refer to children under the age of 12, and this period is a critical period for the growth in life. The design of children's products directly affects

---

① Toys R Us (English: Toys "R" Us, written in its logo as Toys "Я" Us) is the world's largest retailer of toys and baby products. It integrates various brands to provide consumers with a comprehensive, one-stop shopping experience.

children's physical and mental health. Children at this stage have their own preferences and cognition. For example, their preferences for colors, shapes, and cartoon characters all affect the concept of the design of children's products. China is a country with a large population. Children under 12 account for about 25% of the total population. Although the children's product market is growing rapidly, the following problems still exist in the current development of the industry: ① There are many children's product manufacturers, but their market share is low. At present, China's children's product manufacturing industry is still dominated by production rather than design and research, and the manufacturing technology level of children's products is mainly concentrated in the low-end level. Compared with the level of R&D and design of well-known international brand companies, Chinese companies still have a certain gap, and the influence of domestic brands is low. It is precisely because of the backwardness of R&D and design level, brand influence, quality level, etc., that domestic children's product manufacturers are small and concentrated, and the level of production manufacturing is low, mostly OEM low-end manufacturing. ② The regional chain of products is relatively concentrated, and most of the product import and export bases are concentrated in Suzhou, Jiangsu, Dongguan, Guangdong, Wuzhen, Zhejiang, Qingdao, Shandong, Xiamen, Fujian, etc. According to the report statistics, in 2015, the export value of children's products nationwide was 30.803 billion U.S. dollars, and the industrial cluster area totaled 28.987 billion U.S. dollars, accounting for 94.10% of all exports. ③ OEM (Original Equipment Manufactures)[①] is the main mode, and multiple business methods coexist. In the OEM model, children's product manufacturing companies provide brand companies with pure processing and manufacturing services. Because the added value of processing and manufacturing is very low, the profit of the product is very meager. At present, some domestic children's production companies have begun to build OBM (Original Brand Manufactures)[②] of their own brands. Compared with the OEM model, the self-developed brand has a higher technological content, and the profit margin of the brand is also greater, but the building of brand influence takes time. ④ Children's product manufacturers are constantly seeking transformation, and their integration with the cultural industry has been continuously strengthened. On the one hand, the integration with the animation industry has been continuously strengthened, such as integrating products with the animation industry and innovating, turning classic images in animation into dolls and

---

① The basic meaning of OEM (Original Equipment Manufacturer) is manufacturing cooperation with fixed brands, commonly known as "manufacturing for others". OEM products are customized for brand manufacturers. After production, the product can only use the name of the brand, and absolutely cannot be produced under the name of the manufacturer. The characteristics of OEM products are: technologies are outside, the capital is outside, the market is outside, and only the production is inside. Therefore, the quality of OEM products may not be the same as the quality of original products.

② OBM refers to the OEM manufacturer operating its own brand. OEMs need to have a complete marketing network to support OBM. The cost of channel construction is very high, and the energy spent is much higher than that of OEM and ODM, and they often conflict with their own OEM and ODM customers. Generally, in order to ensure the interests of major customers, foundries rarely do operate OBM with a big fanfare.

giving them vivid interactive stories. On the other hand, children's product manufacturers continue to enter the Internet education and entertainment market. The vigorous development of the Internet, CG animation, and VR/AR technology has promoted the rapid rise of online education platforms. For example, the domestic grape technology company provides a variety of ecological games and learning products for children aged 3-12, and creates various imaginative products for children through advanced technology. For example, the children's Q Tao logic party uses logic training and AR teaching to help children learn basic colors, graphics recognition and spatial orientation, length comparison and other knowledge. The natural entertainment attributes of children's products have enabled more and more interesting and innovative products to be developed. They have both vivid stories and cool entertainment functions. The transformation and upgrading of children's products provides a rare opportunity for accelerating industrial development.

2.4.1　Types of children's products

The dazzling array of products in the children's product market shows people's diversified needs for products. Every children's product has many integrated characteristics. This book mainly divides the design of children's products into two categories according to the application mode of interactive technology, one is traditional children's products, and the other is interactive smart products.

Chapter 2 Theory of the Interaction Design of Children's Products

1. Traditional children's products can be divided into entertainment products, educational products, and nursing products according to their functional types and user needs.

① Entertainment products mainly use entertainment and games to cultivate children's interesting and interactive game experience, exercise children's good psychological quality, patience, carefulness, self-confidence and reverse thinking, and exercise children's hand-eye coordination ability. For example, building blocks can cultivate the ability of spatial reasoning, observation, and cognition of the structure of objects. For example, observing the appearance characteristics of playground facilities, landscape environment, characters, etc., and then using building blocks to build them. Repeatedly building various objects not only improves children's observation and memory skills, but also cultivates their operational and creative abilities.

② Educational products mainly refer to teaching aid products that assist children in learning. For example, Rubik's cube or maze toys, teaching aids can improve children's memory, overall observation ability and spatial thinking ability. When children play Rubik's Cube, their hands are playing, their eyes are watching, and their brains are always wondering how to play it. Through games, children can learn, feel, experiment and master knowledge. The brain also absorbs new experience and knowledge in this way. Children's mind and body are exercised and developed unconsciously.

③ Nursing products mainly refer to children's products that support children's health and safety needs. For example, safety seats, tableware, etc., are specially designed for children's special stages of physical development and learning and mastering life skills. Traditional children's products are ingeniously designed and are essential toys for every child's growth. However, in the smart age of "technology +", the interaction between traditional products and children is too single, which will make them feel bored. They can't keep children interested for a long time, and can only meet the basic functions, as shown in Table 2.1.

Table 2.1 Classification of traditional toys (drawn by the author)

Basic Information	Functional Description	Strengths Describe	Defect Description
Building blocks	Designed for the small children of large particles products, easy to insert and disassemble, basic, fire station, zoo and other sets, the shape is very cute	It can improve children's observation and memory, as well as their ability to operate and create	In the intelligent era of "science and technology +", the interaction between traditional products and children is too simple, which will make children feel boring and can not make children interested for a long time, but only meet the basic functions and functions
Children's rubik's cube	Made of flexible hard plastic. The idea is to disrupt the imitation and then recover it in as little time as possible	Children can learn, feel, experiment and master knowledge. The brain absorbs new experiences and knowledge in this way, and the child gets physical and mental exercise and development	
Children's tableware	Children's tableware material is generally chosen with light quality, not easy to fall, environmental protection and health. The shape is round and lovely, bright color increases the child's movement and stimulates the nerve to make the child's brain develop as soon as possible	It is designed for the special stage of children's physical development, learning and mastering relevant life skills	

Chapter 2　Theory of the Interaction Design of Children's Products

2. Nowadays, there are more and more interactive children's products[①] in the market. Smart children's products are developed on the basis of traditional products, and are designed with scientific and technological content to make smart children's products more humanized, intelligent, and technological.

From the analysis of the current market situation of interactive children's products, the market share of smart children's products is increasing. According to the interactive mode of smart children's products, they can be divided into three categories: body touch interaction, voice control interaction, image interaction, etc.

① **Products with body touch interaction.** Body touch interaction is a common way of interaction, which uses body touch interaction to realize human-computer interaction on physical products or touch screens. Once a child touches the sensor interaction module inside the product with his hand or other body parts, it will produce mechanical vibration or sound. The different interaction behaviors of the body and the sensing of different positions of the sensor module will make the product give different effective feedbacks and increase children's interest in the product. As shown

---

① Interactive children's products are a market segment of the product category. They integrate IT technologies with traditional products. They are a type of smart products different from traditional children's products. They have gradually become popular in recent years. Because they are new things, there is currently no industry standard and no authoritative organizations have given them a complete definition. If we summarize the common features of most interactive children's products in the market to give a definition of smart children's products, the definition is: products with cartoon shapes, voice interaction, or simple interaction with people. Smart products have integrated plush toys, rubber dolls, chips, digital technology and other products in different industries, which can achieve a strong effect of edutainment.

in Table 2.2, the Ziro smart glove uses gestures to control the building block robot, which is equipped with a flexible bending sensor, which can recognize and respond to 7 kinds of gestures. This greatly enhances the user's control and gives children a different experience; the Phiro smart car can be controlled in a variety of ways. Lego blocks are assembled to form a variety of toys, and it has multiple functions such as sound playback, face recognition, and image capture; in the Robo Wunderkind robot building block combination, each smart building block has an independent functional module. Through the free combination of smart building blocks, different interactive effects can be achieved, and the building blocks can be controlled by mobile phones to achieve different functions. The modules are spliced together and operated on the app through Bluetooth and mobile phone connections.

Table 2.2 Products with body touch interaction (drawn by the author)

Product name	Advantages	Disadvantages
Ziro Smart gloves	It can realize the recognition and response to 7 kinds of gestures, which greatly improves the user's control.	The function is complicated for young children.
Phiro smart car	It can be controlled in a variety of ways and can be matched with Lego toys to help children learn to program.	Too few styles, not in line with the girl's preferences.
Robo Wunderkind machine blocks	It can be assembled freely, one module has one function, and various functions can be realized through Bluetooth and app.	The operation method of mobile phone or ipad is not suitable for children to play for a long time.

Chapter 2 Theory of the Interaction Design of Children's Products

② **Products with voice control interaction**. Voice recognition meets the interactive needs of children and better improves the educational and entertainment functions of children's products. A large number of innovative interactive methods such as voice control and physical touch have been favored by children and parents. The use of innovative voice dialogue design in children's products will greatly enhance children's interactive experience, improve the short duration of children's interest in products to a large extent, and strengthen children's language expression ability. As shown in Table 2.3, the Hummingbird Robotics kits paper robot teaches children to make their own robots through simple micro-sensing device manufacturing principles, and controls the movement of the paper robot through voice dialogue. Children can use paper plates to make a variety of creative shapes, and it also incorporates traditional manual painting techniques into robot making; KOMO educational robot adopts NFC card[①] + voice interaction, allowing the robot to read two-dimensional plane pictures by swiping the card, and display them in three-dimensional way. The content of the voice dialogue mainly includes knowledge questions, poems, calculations, weather, encyclopedia, etc. By touching the head of the robot, children can get different interactive feedback.

---

① The full English name of NFC is Near Field Communication, which is a wireless technology initiated by Philips and jointly promoted by famous manufacturers such as Nokia and Sony. Not long ago, a number of companies, universities, and users jointly established the Pan-European Alliance to develop an open architecture for NFC and promote its application in mobile phones. NFC evolved from the integration of RFID and interconnection technologies. It combines inductive card reader, inductive card and point-to-point functions on a single chip, enabling it to identify and exchange data with compatible devices within a short distance. This technology was originally just a simple merger of RFID technology and network technology, and now it has evolved into a short-range wireless communication technology, and its development is quite rapid.

Table 2.3 Products with voice control interaction (drawn by the author)

Product name	Advantages	Disadvantages
Hummingbird Robotics kits paper robot	Combining painting and handmade elements, giving children endless space for artistic design .	The interaction mode is relatively simple, and the material is easily damaged.
KOMO Education robot	Recognizing and reading two-dimensional pictures, and displaying them in a three-dimensional way. The content of voice interactive dialogue is diversified.	The location of the NFC is not clearly marked, which is obviously not intuitive enough for children who use it for the first time.

③ **Products with image interaction**. Smart products based on image interaction can promote the training of children's perception and logical thinking abilities, cultivate children's ability to recognize the colors and graphics of things, and improve the hand-eye coordination ability of using hands and brains. For example, Weidou Solitaire is an early childhood education game Solitaire, an innovative children's game that combines physical card products and APP. In play-based learning, children need to manipulate props such as cards on paper entities. And interact with the APP controlled by the parents. Parents can record their children's reactions on the APP, and they can also tell the children the stories in the game to enhance parent-child companionship; "Mystery Mirror Magic Painting" is a drawing virtual reality game. It uses white paper to paint freely, and expands its imagination through auxiliary materials. After scanning the underwater world drawing paper, the screen will automatically recognize it, enter the 3D ocean scene, and draw small fish and water plants on the paper. It also provides small cards to help children identify underwater creatures; "LEGO Fusion" is a

Chapter 2 Theory of the Interaction Design of Children's Products

combination of Lego bricks and a special base plate, and is used in conjunction with the IOS application. After building the castle, take a photo with an iPad, recognize the building blocks through Qualcomm's viforia AR image technology, and input the physical castle into the iPad to become a virtual castle, as shown in Table 2.4.

Table 2.4 Products with image interaction (drawn by the author)

Product name	Advantages	Disadvantages
Weidou paper cards	It focuses on parent-child interaction and companionship, and helps children learn simple calculations, graphics, and vocabularies through tangible interaction media	Not smart enough, paper products are slightly "simplistic".
Mystery Mirror Magic Painting	It can freely paint on white paper and expand imagination through auxiliary materials.	Painting is limited to fixed themes.
LEGO Fusion	The physical castle is input into the iPad to become a virtual castle.	Picture recognition, no nfc, no serial number authentications.

Through the comparative analysis of the market of traditional products and interactive smart products, the researchers believe that the current children's products are developing in the direction of intelligence, fun, and multi-function. The design should fully tap the advantages of traditional children's products with simple structure but strong variability, and apply them to the design and development of smart products. Products aimed at children should not only avoid the frustration of the complexity and cumbersomeness of modern technology, but also enhance the fun of children's interaction with the help of advanced interactive technology, so as to make use of the features of interactive smart products that are entertaining, interactive, learnable, upgradeable, etc., to build a new interactive children's product that complements intelligence and tradition.

2.4.2　Development trends of children's products

The market segmentation of children's products is obvious. Because children of different ages have significant differences in physical, psychological, cognitive, and game development abilities, the design of children's products needs to be adjusted differently based on children's cognitive behavior characteristics. For example, pre-school children between the ages of 3 and 7 prefer to imitate the words and actions of others, like figurative cartoon toys, express personal emotions independently, and have a strong dependence on toys, and begin to become interested in smart products. Products designed for children should meet the physical and psychological cognitive needs of children, and designers need to understand the current situation of the children's industry. At present, the development trend of the global children's industry mainly has the characteristics of entertaining and learning, situational interaction, new technology, and high safety.

1. Comprehensive development of product categories.

① The types of children's products have broken through the age limit. With the development of new interactive technologies, the functions of smart products have become diversified. This not only provides entertainment restrictions for children, but

Chapter 2 Theory of the Interaction Design of Children's Products

the development trend of children's products is also becoming more technological and intelligent. More and more parents are willing to consume high-tech products to play with their children. ②The integration of traditional products and smart products. Technology has penetrated deeper and deeper into the children's industry. AR/VR technology, voice interaction technology, face recognition technology①, brain wave induction technology②, etc., are deeply integrated into children's design. Well-known global brands have launched smart children's products. For example, Sony launched its robot product "toio" at the Tokyo Toy Fair in 2017. Children can place various shapes they set on the "smart module", and then use the ring handle to control the robot to move back and forth, left and right, and rotate. Children can cooperate and play against each other. However, the rise of smart children's products has not had a serious impact on traditional children's products. Traditional children's products can give children the space to exert their imagination, and the design of the shape is also in line with children's preferences. But they are less entertaining. Therefore, traditional children's products must also keep pace with the times and combine the fun, interactive, environmentally friendly and green advantages of most smart products to promote parent-child interaction while evoking childhood memories. For example, the famous

---

① Face recognition technology refers to the use of computer technology for analysis and comparison to recognize faces. Face recognition is a popular computer technology research field, including face tracking and detection, automatic adjustment and enlargement of images, night infrared detection, automatic adjustment of exposure intensity and other technologies. Face recognition technology is a kind of biometric recognition technology, which distinguishes individual organisms through the biological characteristics of the organism (generally referred to as people).

② A brain wave sensor is an electronic device that uses human brain waves for manipulation.

children's brand Lego launched the STEM series of suits "Brainstorm." It allows children to use traditional building blocks to make Lego robotic creatures, vehicles and other inventions. The building block parts can be combined with motors and sensors, allowing the robot to walk, talk, grab, think, etc. The future trend of the global children's industry will be the integration and coordinated development of traditional products and smart products.

③ The linkage with the cultural industry is strengthened. The prosperity of the cultural content industry such as animation and picture books has provided a lot of materials for the development and design of children's products, and broadened the research and development ideas. Adding cultural elements to the design of children's products can enhance the added value of the products and enhance consumers' loyalty and recognition of brand products. In particular, peripheral cultural and creative products derived from popular animation works, such as classic stories, have cultural elements such as characters, stories, and fun. The prototypes of the hot-selling Disney series, Transformers, Ultraman, My Little Pony, Super Flying Man and other dolls in the market are all derived from related film and animation works. The combination of children's products and the cultural industry has created a good brand image and enhanced the brand's popularity and reputation. Animation peripheral products are quite popular among children.

2. Safety standards in the children's product industry are becoming stricter.

The National Standards Committee and the General Administration of Quality Supervision, Inspection and Quarantine issued GB 31701-2015 *Safety Technical*

Specifications for Infant and Children Textile Products, and the new *National Toy Safety Technical Specifications* was formally implemented in 2016. At present, the characteristics of children's product safety standards are as follows: ① The safety standards increase the scope of protection for children, including providing safe manufacturing standards for products and materials for children under 14 years of age. They also include products that are not toys but have play functions for children under 14 years of age. On the other hand, safety standards define strict standards for the sound, mechanical parts, power supply and other safety indicators of children's products; the new safety standard embodies the principle of participation in line with international standards. The new safety standards are in line with the "International Children's Products and Toys Safety Standards" (ISO 8124), while referring to strict EU safety standards, and the core safety technical standards are consistent with international standards. When designing children's products, designers should mainly consider the safety of children's products from the following aspects: The safety of the shape of children's products. Products for young children should use rounded and smooth corners. Materials with large gaps or uneven surfaces should not be used. The details of the product should be considered, and the children should not be caught or hurt. For example, the mushroom nail box puzzle is made up of colorful patterns with mushroom-shaped small particles of various colors and sizes. However, the building block particles are too small, and the sharp corners at the lower end are like small thumbtacks, which can easily cause problems such as swallowing and difficult grasping for children. Children's body organs, bones, skin, etc., are very immature, and their self-protection awareness is poor. Product designers should not use sharp shapes, edges or unsuitable gaps when designing children's products. ② Safety of materials. Children's products are one of children's closest partners. Not only do they have skin contact, but young infants and young children often bite and kiss products. The quality and safety of product materials and the materials used are issued that need to be considered in the design of children's products. With the development of information technology, material technology has become more and more high-tech. Materials with various characteristics are used in children's products. For example, magnetic fluid toys that have both liquid-like fluidity and solid magnetic materials can become ink-like powders when shaken, and when they encounter magnetism, they immediately become arrays. Of course, when selecting materials, one should not only pursue appearance and function, but ignore the dangers of the material itself. ③ The structure rationality of children's products. Children's products have relatively high structural requirements. For example, plastic toy airplanes contain small parts, which are easy to fall off from the airplane. If they are swallowed by children, they may cause choking hazards. Children's products are often smashed and parts are easy to fall off. When designing children's products, the structure can be simple, but it must be firm to avoid scattering and causing secondary damage.

3. Children's products are increasingly interactive.

The "interactivity" of children's products means that children interact with people around them through the product and can achieve certain functions. Good interactive products can bring more fun experience to parents and children. For example, WooWee

Chapter 2   Theory of the Interaction Design of Children's Products

Fingertip Monkey is a robot monkey that interacts with people. It has a variety of play actions and imitated sounds, allowing people to understand their feelings. Children can shake, caress, kiss, and even coax them to sleep. The Barbie DreamHorse manufactured by Mattel can interact with people flexibly and respond to people's touch and sound. When the horse's body is touched, it will walk up and move in the direction of the sound. For example, if you make a sound on the left side of the horse, it will go in that direction and feed it plastic carrots. The little pony will make a sound of eating, and the little pony will dance to music. When asked a question, it will nod (as shown in Fig.2.27b). These products use sensor technology to make children feel the magical interactive experience fun. It is these dolls with human emotions that give children emotional companionship and subtly master the energy of high-tech, so that they can better adapt to the needs of the rapid development of the information technology era.

4. Importance is attached to the development of domestic brands.

Brand creativity is the core competitive element of the children's industry. Domestic children's product companies used to mainly process and OEM production, and could not have their own core competitive products. Due to the limitations and dependence of the OEM model, some enterprises in the children's industry and Internet technology companies have begun to pay attention to the construction of their own brands, increase investment in scientific and technological strength, optimize channel layout, and introduce cultural and creative connotations, thus establishing domestic brands and realizing independent operation at the same time. For example, the well-known grape technology company in China adopts international advanced teaching and education concepts and uses high-end information technology to create various imaginative products for children, bringing better parent-child companionship. The company's "AR Photo Gallery" uses AR to enhance the painting experience, free setting mode, story-based coloring book, with colorful illustration color matching, and hand-made DIY photo frame display, making painting more fun. Beijing Xiaoxiaoniu Creative Technology Co. Ltd., is a high-tech enterprise focusing on children's educational education. It uses natural human-computer interaction technology to provide children with creative and fun open educational and entertainment experiences. "Mystery Mirror Magic Pen" takes the magic pen Ma Liang as the creative prototype, and integrates AR technology into the painting, and recognizes children's paintings or plasticine handmade works through the mobile phone camera. In this way, it truly achieves "superiority", stimulates children's imagination and creativity, and makes them become "magic brush Ma Liang".

2.4.3　Design elements of children's products

The consumers in the children's product market are parents, while the direct users are children. From the perspective of the needs of children's parents, the design of children's products requires not only a certain educational function for the children, but also an interactive parent-child entertainment function. At the same time, the safety of the product must also be considered. On the other hand, from the perspective of children, designers must follow the principle of everything for children. In the design of children's products, they must attract children's attention. For example, the shape is

Chapter 2 Theory of the Interaction Design of Children's Products

lively, cute and full of fun, colorful, and versatile. It can also be combined with children's familiar animation characters in the game content to attract children's interest and enhance their love for the products.

1. Elements of artistic forms.

Art form refers to the external form of the user's senses, which is determined by various media that shape the artistic image. The artistic form of the design of children's products mainly refers to shape, color, material, etc. ① Round and lovely design. Children's favorite cartoons and animations magnify a certain feature in an exaggerated way. For example, the big-bellied Xiong Da and Xiong Er deliberately exaggerated the bear's belly in order to emphasize the innocent image. In the design of children's pound, the rounded and lovely shape not only meets the physiological safety needs of young children, but also satisfies their preference for shape recognition. For example, the "Dash & Dot" programmable robot has a chubby and cute appearance, and is equipped with four advanced apps for children of different ages. The sleek shape allows the robot to follow the route or play music, and the full shape ensures that children are not likely to cause physical injury when playing. ② Bright and lively colors. Brightly colored objects can more attract children's attention and are deeply loved by children. At the same time, colorful and lively products can stimulate children to learn color associations. For example, when they see red, they will think of flames, when they see blue, they will think of the sea, and when they see green, they will think of the grass and when seeing orange, they think of oranges. These colors represent positive energy and cater to children's psychology. ③ Material elements. In the design, the material technology itself is an important factor that constitutes the aesthetics of the product. How to reasonably apply the advantages of material technology to the design of children's products is the embodiment of modern high-level industrial production technology and modern aesthetic concepts. Different materials give people different psychological perceptions. Plastic gives people a warm and smooth psychological feeling; metal gives people a heavy and hard psychological feeling; plexiglass gives people a translucent, shiny and clear psychological feeling; wood gives people a feeling of lightness, simplicity and warmth. In the design of children's products, various factors such as the safety of materials, the beauty of texture, and the new technology must be fully considered.

2. Intelligence element.

With the rapid development of information technology, the product structure of the children's product market has changed. In particular, the domestic children's product market has gradually developed from low-end products with traditional styles, single functions, and no differences to modern, multi-functional, and intelligent trends. The famous psychologist Gardner put forward the theory of multiple intelligences[①].

---

① Gardner's Theory of Multiple Intelligences. The traditional theory of intelligence believes that human cognition is unitary, and that individual intelligence is single and quantifiable. But the American educator and psychologist Howard Gardner proposed in the book "The Structure of Intelligence" published in 1983 that "Intelligence is the ability of individuals to solve real problems encountered by themselves or to produce and create effective products under the value standard of a certain social or cultural environment or cultural environment". Everyone has at least language intelligence, logic and mathematics intelligence, music intelligence, spatial intelligence, body movement intelligence, interpersonal intelligence and introspective intelligence. Later, Gardner added natural intelligence. This theory is called the theory of multiple intelligences.

Chapter 2 Theory of the Interaction Design of Children's Products

From a scientific point of view, it finds out the characteristics of children's eight intelligences (such as language, logic, space, movement, etc.). The development of each kind of intelligence has its own unique sequence, which matures in different periods of the children. With the penetration of the "STEAM" concept in the field of children's education, smart products with elements of science and education can teach children to learn programming and advanced educational concepts while playing. For example, Fisher-Price launched a "Think & Learn Smart Cycle", which uses a Bluetooth bicycle to quickly pair with a tablet or streaming TV device. The player can control the progress of the game through the motion of the pedal. The educational purpose of the smart bike game is to let children know new words by choosing letters. And after the game, children can review the new words, so as to achieve the development of motor, language and natural observation intelligence. Another social robot called "Leka" is designed for children with autism, Down syndrome, or multiple disabilities. Leka provides sensory stimulation to help children with special needs become more independent and improve their athletic and social skills. Leka smart products and companion apps can provide interactive educational games for children with special needs. The coolest feature of Leka smart products is that the difficulty of each game can be customized to ensure the learning of children in each game. The impact of mobile internet and multimedia content on the children's industry is increasing. The trend of intelligent and interactive development of children's products is obvious. The design of children's productsers should lead the trend and development of products, use the Internet, mobile apps, sensors, etc., to achieve new and natural interactive gameplay, combined with content IP, to increase the product's sense of substitution and attractiveness.

3. Safety element.

The safety of children's products is directly related to the physical and mental health of children. However, due to the immature physical development of children and poor awareness of safety precautions, the body and mind are easily harmed by the external environment. When most children interact with products, they touch, fiddle with, and beat products according to their own behavior. Their self-control ability is poor, and some children's products contain potentially dangerous parts such as plastic, metal, and rope. Therefore, the safety of children's products is particularly important. This is the general trend of the development of the children's product market. ①The product structure should be reasonable. The safety of children's products must be taken seriously in design. This requires the internal structure of the product to be firmed and reasonable to avoid accidental eating or other safety injuries by children due to the falling of small parts. For example, products with internal transmission devices, the closed structure should be considered when designing the product structure, and the dangerous parts should be ensured that fingers cannot reach into the gap. ② Round shapes. The surface of the product should not have sharp edges and corners or excessive gaps to prevent children from being injured when they interact with the product. Similarly, children's products should not be too small to prevent children from accidentally eating. For example, the surface of the product should be painted

Chapter 2 Theory of the Interaction Design of Children's Products

as a whole, so that the whole product gives a feeling of smoothness, stability, comfort and safety. ③ Safety of materials. The use of non-environmentally friendly materials and materials containing heavy metals will cause harm to children's physical health. Non-toxic and harmless materials should be selected for children's products. For example, the coating on the surface of the products should be safe and non-toxic, especially the content of trace elements such as lead and mercury must meet safety standards. For example, the material of the Grape Discovery Smart Cube is food contact grade silicone, which is with a good touching feel, smoothness, no lag, and high drop resistance.

4. Edutainment element.

Learning while playing is the best learning state. Educational and entertaining products are different from other children's products. The most important point is intelligence. When children interact with such products, their intellectual development is subtly affected by the products. Therefore, when designing for children, the design of children's producers should give full consideration to the perfect combination of entertainment and education of the products: This is also the meeting point between children and their parents and meets the common needs in parent-child relationships. For example, the most popular "Learning Resources" gear in the United States is not only a set of flat wood, but also allows children to build freely in a three-dimensional space like an architect. The child is like a mechanic, able to turn all the built blocks. During the whole construction process, children need to use their brains to ensure the linkage of gears. In the process of continuous experimentation, children understand the scientific concepts of geometry, engineering, science, physics, machinery, linkage, and clockwise. Boring technological knowledge is transformed into fun and not boring games. This captures the curiosity of children very well and encourages them to understand how the world works. This is a great set of STEM products. The design of children's products needs to understand children's intelligence level and cognitive characteristics, and at the same time be familiar with the preferences of children's products. For example, children's favorite products are sensory and easy to operate. Products that are too complex sometimes make children feel frustrated, while products that are too simple can easily make children lose interest.

2.5 Summary

This chapter is the theoretical basis of this book. It sorts out and analyzes some of the main theoretical foundations, and conducts an in-depth analysis and sorting of them. This chapter mainly elaborates the theory from three aspects, the first is the theoretical basis of tangible interaction. It traces the principles, characteristics and design elements of tangible interaction, and conducts an in-depth analysis of the theories and cases of tangible interaction. Secondly, it systematically expounds the basic characteristics and interactive behaviors of children's psychology, physiology, and cognition from the perspective of children's psychology, cognition and other professional theoretical disciplines. Finally, it analyzes related theories and methods focusing on the design of children's products, and conducts an in-depth analysis and combing of the product types, development trends, and design elements of the design of children's products. The theoretical basic research in this chapter lays the foundation for the user research in the following part.

Chapter 3 User Research in the Interaction Design of Children's Products

Chapter 3 User Research in the Interaction Design of Children's Products

3.1 Introduction

This chapter is the research and analysis of big data of children groups based on tangible interaction products. It first makes a detailed analysis of children users, comprehensively collects the data of children's development research, and takes young children as an example to establish Kidsplay, a data platform for children's development research from the aspects of physiology, psychology, cognition and game development data analysis. The experimental research data collection and investigation and analysis in this chapter provide a lot of rich user model informations for the design insight, and lay a foundation for the design of the data platform of children's development research. The following chapters 4 and 5 respectively quote the design research methods and conclusions of this chapter, and summarize and sublimate the design theory of children's products. The design practice of children's products in Chapter 6 directly extracts children's physiology, cognitive psychology, game development and other data from the children development research data platform as the design basis.

From the discussion in Chapter 2, we can see that the development of interactive technology has brought new opportunities to traditional the design of children's products. Major technology companies have also tried to use advanced algorithms to optimize all aspects of children's products. But the biggest problem faced by the design of children's producers is the lack of understanding of children's needs. Especially young children are often unable to accurately express their own needs. They also have obvious differences in perception, motor control[①] abilities, cognitive and intellectual levels. Therefore, seeking to provide solutions for the design of children's products at the level of tangible interaction design methodology is the essence of the research problem.

The main part of this chapter discusses the collection and analysis of children's data, and the establishment of a data platform for children's development research. This chapter is organized as follows: Section 3.2 is about data collection and analysis for children. It uses user research methods such as in-depth interviews, questionnaire surveys and experimental observations to give a user model that meets the characteristics of children's profiling. Section 3.3 is the child development research data platform—Kidsplay. It establishes a child development research data platform based on the collection and analysis of children's data, and records the process of platform establishment and the final system model presented. Finally, Section 3.4 is the summary of this chapter, summarizing the research results and innovations of this chapter.

---

① Motion Control (MC) is a branch of automation. It uses some equipment commonly known as servo mechanisms such as hydraulic pumps, linear actuators or motors to control the position or speed of the machine. The application of motion control in the field of robots and CNC machine tools is more complicated than the application in special machines, because the latter has a simpler motion form and is usually called general motion control (GMC).

3.2 Collection and analysis of children's data

The essence of design is "people-oriented" and user-centric. The design of children's products is not only to meet the simple product form, but also to design interactive products that meet the different needs of children of different ages. By observing and capturing children's psychological activities and behavioral characteristics, designers explore the potential physical and psychological needs of children groups to guide the development and design of new products. The characteristics of children's games are completely different from those of adults. This chapter mainly uses research methods such as questionnaire surveys, in-depth interviews, and experimental observation methods to explore the characteristics of children's needs and behavior in physical interactive games. In this way, it sorts out the design data suitable for children's interactive products from different dimensions, established the corresponding data analysis platform and children's role model, so that children can get a better experience in gaming, so as to achieve better edutainment effect.

3.2.1 Interview and questionnaire survey

1. Interview design.

Interview is a systematic and planned way for researchers to collect data through face-to-face interview or group conversation according to the interview outline on the basis of the requirements and purposes determined by the design and research. Children interviews are a way of directly collecting user data. The purpose of this interview is to understand the perception ability, cognitive level, and intelligence level of children of this age.

Interviewees: 60 children aged 4-6 from the Huilongguan campus of the Experimental Kindergarten of Beijing Normal University.

Interview method: The duration of each interview is 2-4 hours, the duration of a single interview is about 20 minutes, the frequency is 2 times/week, and the total duration is four weeks. The recording method is video, records, etc.

Interview content: Based on young children's ability to perceive sets, concepts of numbers, spatial orientation, merging and sorting, symbolic representation and other early mathematical understandings in EMDK (Early mathematics diagnostic kit), it understood children's perception ability and intelligence level. The outline of the interview is shown in Table 3.1:

Table 3.1 Outline of the interview with children (drawn by the author)

No.	Question
Q1	Child's name and age
Q2	What are the colors in the picture, please? (Identify more than 2 colors)
Q3	Please tell the shapes in the picture. (Identify more than 2 shapes)
Q4	Please name the things in the picture. (Point to the carrot on the right and the rabbit on the left respectively)

continued

No.	Question
Q5	If you have so many rabbits in the picture, are these carrots enough for each rabbit to have one carrot?
Q6	Which car in the picture ranks first? Which one is last in the row? Which one comes second?
Q7	How many apples are there in the picture? How many are left after eating two?
Q8	In the picture, which of the children is tall and which is short?
Q9	Please take a look at the story of the four pictures and put the pictures in the order in which they happened.
Q10	Please take a look at the statistics of the little animals that children like in the picture. How many children like dogs? Which kind of animal do children like most?
Q11	Please tell us how many of these are in the picture? Please write it down.

Aiming at the cognitive abilities of pre-school children, the researchers drew corresponding cognitive cards and used them in conjunction with the interview outline in the list.

Summary of the interview: To understand children, the first step is to contact them and observe cognitive behavioral characteristics. Through single-person interviews with pre-school children, the researcher has a preliminary understanding of the characteristics of children at this age, laying a foundation for further research on children's play preferences. The researchers analyzed and selected the content of the interviews in terms of age, sensory cognitive characteristics, etc. as follows: ① Children aged 3-4 have strong perception, can recognize colors and graphics, and their mathematical ability is within 10. They are eager to see, hear, touch, and smell information, understand the meaning of words and pictures, and like to draw without logic, and use pictures to express meanings. ② Children aged 4-5 have more proficient action skills, are more sensitive to colors and graphics, and their mathematical ability is within 50. They like to imitate the behavior of adults, are gender-conscious, can tell stories based on continuous pictures, and like to use pictures and symbols to express their wishes and ideas. ③ Children aged 5-6 have the ability to understand symbolic language, their ability to recognize colors and graphics is strong, and their mathematics ability is within 50-100. They have flexible movement skills, can accurately express their opinions, and like to use pictorial symbols to express things and stories (as shown in Table 3.2).

Table 3.2 Summary of interviews with children (drawn by the author)

Age	Characteristics	Reading habit	Comprehension	Written expression
3-4 years old	Strong perception, eager to see, hear, touch and smell information, curious about everything	They will take the initiative to ask adults to tell stories, they like rhythmic nursery rhymes, and they like to watch cartoon videos	They understand short nursery rhymes and stories, can read pictures and speak, understand the meanings of words and pictures in books	They like to paint without logic, and use pictures to express meanings
4-5 years old	They have more proficient action skills, like to imitate adult activities, and are gender-conscious	They will read the content they like repeatedly, they like to tell stories to others, and they know the meanings of the symbols	They can tell roughly what they have heard, they can tell stories based on continuous pictures, and they can express emotions such as joy and worry along with the stories.	They like to use pictures and symbols to express their wishes and ideas. Adults remind them to maintain the correct posture

Chapter 3　User Research in the Interaction Design of Children's Products

continued

Age	Characteristics	Reading habit	Comprehension	Written expression
5-6 years old	They can understand sign language, are not good at reading, have flexible movement skills, and are rebellious	They can concentrate on reading, like to communicate the content of the story with others, are interested in books and words, and know the meaning	They can tell the main content, guess the progress of the story, express their opinions, and feel the beauty of language	They like to use pictures and symbols to express things and stories, can write their names basically correct, and write and draw in the correct posture

2. Questionnaire design.

Questionnaire survey is to collect data from a large number of people, including users' opinions, attitudes, preferences, personal information, etc. (either abstract concepts or specific habits or behaviors), so as to obtain relevant data and mine information related to product design, user interface and usability. This questionnaire survey was based on the interview framework and adopts a semi-structured questionnaire. The researcher filled in the questionnaire based on the children's answers. The purpose of the questionnaire survey is to further understand pre-school children's personal preferences for products and children's gaming habits and other information.

Respondents of the questionnaire survey: A total of 109 children aged 3-7, including those from the Experimental Kindergarten Affiliated to Beijing Normal University Huilongguan Campus, Beijing Fanghua Kindergarten in Red, Yellow and

Blue, and shopping malls in Beijing; among them, 59 are boys and 50 are girls.

Questionnaire method: The questionnaire is a semi-structured questionnaire. The researcher uses face-to-face communication with young children and records various responses of the respondents in accordance with the format and requirements of the questionnaire. The duration of each survey is 2-4 hours, the duration of a single interview is 20 minutes, and the recording method is transcript.

Information available: 109 valid questionnaires were recovered, and the numerical value of the selected questions and the content of the open question and answer questions were in line with the standards of quantitative and qualitative analysis.

Questionnaire content: ① Design basis. The questionnaire design is based on the theoretical analysis and interviews in the previous chapters. The purpose of the research is to understand the product preferences of pre-school children and factors such as their cognition and preferences for games. The design of the questionnaire options sent to pre-school children is intuitive and interesting, because the children's reading level and comprehension ability should be taken into consideration. ② Questionnaire design framework. The questionnaire survey is designed in five aspects: basic information, game behaviors, product preferences, artistic cognition and views on children's products. The design framework is shown in Table 3.3.

Table 3.3 The design framework of children's questionnaire (drawn by the author)

Basic information of the children	Gaming behavior	Product and relevant preference	Art form cognition	Opinions about children's products
Name	Do you like playing games with kids?	Please choose the type of product you like from the following sixteen pictures of toys	Please choose your favorite product shape from the following four types of shapes	What are the features of the product you want the most (choose three items or fill in by yourself)
Age	Please tell me your favorite games	Please choose the products you usually like to play from the following options	Please select the product features that appeal the most to you from the following options	Please design a product. What functions do you want it to have (choose three or fill in by yourself)
Gender	Please tell me who you like to play toys with	Please select from the following options how often you usually play with toys	Please choose your favorite color from the pictures below (choose three more)	Please tell us what entertainment products you currently have? What is your favorite?
Interview time		Where do you usually play games	Please select your favorite graphics from the options (three options can be selected)	Please tell us what your ideal game product is like?
		Please choose which style of building blocks you like	Please choose the product material you like	

continued

Basic information of the children	Gaming behavior	Product and relevant preference	Art form cognition	Opinions about children's products
		Please choose which Lego product you like from the pictures below		
		Please choose how you feel about Lego products (three options can be selected)		

Aiming at the reading level and figurative thinking style of pre-school children, the researchers provided figurative picture options in the options to match the content of the questionnaire outline.

Summary of the questionnaire survey: Through statistical analysis of data, this study quantifies children's behavioral preferences for interactive product games, starting from the exploration of the design of children's products, and the analysis conclusions of children's product surveys for the 3-7 year-old age group are as follows:

① A total of 109 people were surveyed in this questionnaire survey, and the age distribution of children was 3-7 years old. Among them, 4, 5, and 6-year-old children accounted for 68% of the total number, accounting for a relatively large proportion; boys and girls were evenly surveyed; 69% of children like to play games with their peers very much. Children like different types of games, boys like building blocks, cars, video games, sports activities, etc., girls like building blocks, playhouses, dolls, etc. Building blocks are neutral products popular among children (as shown in Figure 3.1).

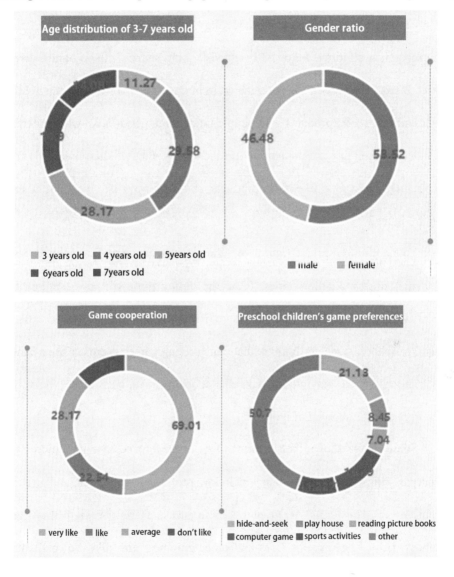

Figure 3.1 Analysis 1 of the questionnaire survey among children (drawn by the author)

② Building blocks and dolls are the favorite entertainment products of young children; children's favorite products are mainly dolls. Young children have a high acceptance of animal modeling products, and children aged 6 and above have a high acceptance of abstract and cartoon modeling; color is the most attractive product feature for young children, followed by good-looking shapes, moving and interesting products. It can be seen that vision is the main interaction channel for young children; younger children prefer products with simple interaction, good-looking and interesting, while older children pay more attention to functions; young children prefer plastic and plush products, most girls like plush products, and most boys like plastic products (as shown in Figure 3.2).

③ Young children prefer figurative style products, such as Lego city series, fire truck theme, creator building series. From the design point of view, children prefer products with good-looking colors and cute shapes that can be shared with their parents. Pre-schoolers generally think that constructing games exercises their hands-on ability, can learn while playing, and can also cooperate with classmates, which makes them very happy (as shown in Figure 3.3).

④ Studies have shown that in terms of color preferences, young children prefer blue, purple, yellow, red, etc., among which boys prefer blue and green, and girls prefer red, purple, etc. In terms of graphics preferences, there are significant gender differences. Boys like firearms, cars and other graphics, girls like flowers, cats and dogs and other graphics. Boys are more likely to accept abstract graphics than girls.

The frequency of young children playing with toys decreases with age. 3-year-old children play with toys every day, while 10% of 7-year-old children have rarely played with toys. Over 60% of children often play games at home, as shown in Figure 3.4.

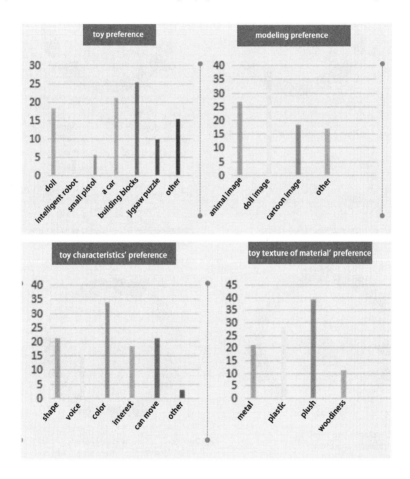

Figure 3.2　Analysis 2 of the questionnaire survey among children (drawn by the author)

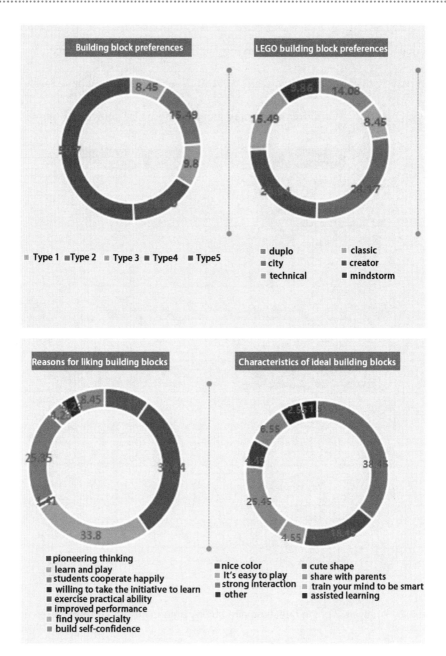

Figure 3.3 Analysis 3 of the questionnaire survey among children (drawn by the author)

Chapter 3　User Research in the Interaction Design of Children's Products

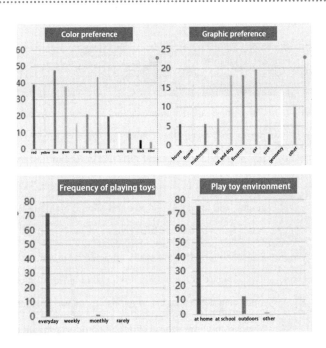

Figure 3.4　Analysis 4 of the questionnaire survey among children (drawn by the author)

According to the analysis of literature theory, interviews and questionnaire surveys, this study has a preliminary generalization and summary of the entertainment and education product preferences and game characteristics of preschool children, as shown in Table 3.4:

Table 3.4 Children's product preferences and game features (drawn by the author)

Summarized characteristic	Detailed product preferences and game features of pre-school children
Building blocks are the most popular neutral products	Building blocks are a favorite product of pre-school children, and are not affected by gender differences, and are a popular neutral product conducive for intelligence development.
Product material should be comfortable and soft	Preschoolers like soft materials. Most children's products are made of plastic and they like auxiliary materials.

continued

Summarized characteristic	Detailed product preferences and game features of pre-school children
Blue, yellow, purple and green are children's favorite colors	The neutral colors that pre-school children prefer are blue, yellow, purple, and green. The graphics are greatly affected by gender. Boys like cars and firearms, and girls like dolls and flowers.
The game situations are mainly concrete	The cognition of pre-school children is concrete. They like situational games, and the design of products should highlight themes, and it is better to have concrete life scenes.
Product interaction should be based on multi-channel design	Pre-school children are easily attracted by products with bright colors, lively shapes, and dynamic products. In terms of interaction methods, multi-channel design based on vision can be considered.
The products create a new gaming experience	The design of pre-school children's products should create a new game experience and be flexible and changeable, so as to stimulate children's interest and curiosity.

The above summary is the game experience characteristics of pre-school children's interactive products summarized through user interviews and questionnaires. However, for the product design of educational building blocks, it is necessary to dig deeper into children's cognitive psychology, emotional experience factors and reasonable design of game content during game play. In the following research, the researchers will use experimental observation methods to further obtain the most realistic emotional characteristics, cognitive characteristics and game development skills of pre-school children in the process of playing educational games.

3.2.2 Experimental observation and analysis of game behavior

Block building games are a type of educational games very popular among pre-school children. This research adopts the experimental observation method to analyze the development trend of children's building block games from the aspects of skill behavior, creativity level, and emotional experience of games built by pre-school children's building blocks. In this way, it understands the characteristics of pre-school children's physical, psychological, cognitive, and emotional experience, providing a theoretical basis for further designing children's interactive products and guiding children's building block games.

Purpose: by observing the building block game process of pre-school children, this experiment aims to obtain more intuitive data of children's cognition. ① To understand the development characteristics of children's physiology, cognitive behavior, and emotional experience in building block games. ② To understand various problems in the process of building blocks for children. ③ To provide theoretical basis for designing children's construction products.

Experimental subjects: 60 children aged 4-6 in Beijing Red, Yellow and Blue Beijing Fanghua Kindergarten; among them, 20 are 4-year-old children and 10 are male and female students; 21 are 6-year-old children, 11 boys and 10 girls.

Experiment procedure: The researchers provided 150 blocks of different shapes. Children were asked to choose a kind of block and build their own home within 15 minutes. The pre-experiment invited 4 children to complete the task within the specified time. In the process of building blocks, the subjects were reminded of the building theme at the 3rd and 8th minutes after the start of the game. ① Pre-observation: pre-observing the process of building blocks by 4 children, observing whether the children's building ability meets the requirements of the proposition within the specified time, and conducting pre-assessment with reference to the children's work with the highest level of construction. ② Formal observation: The author can only bring one subject to the game room at a time and use the building blocks provided by the desktop to build things. Before the start of the game, the author introduced various building blocks to the subjects and displayed reference objects, and asked the children to use existing building blocks to build the proposition. The whole video recording, the longest time is 15 minutes, records the construction skills, construction behaviors, reasons for emotional changes used in the construction process, time to complete the work, number of building blocks, materials, problems encountered, etc. By analyzing each construction process and the photos of the final work, we collected data on the construction level, creativity level, emotional experience, shape combination and other data of the work, and processed and analyzed the data.

Data collected from the experiment: 60 children's video games, photos of their works and paper text materials recorded during the observation process were collected.

Summary of the experiment:

① **Analysis of the development level of children's block building:** This experiment used Casey and Andrews' tool of building block construction level test to evaluate children's level of building block construction. The results of the experiment are shown in Table 3.5 below. The building block level of children of this age range is between 2 and 5 points (full score is 9). That is to say, in the process of building blocks, 4-6 year-old pre-schoolers will use multi-dimensional vertical or horizontal laying, overhead, enclosure, regular enclosure, solid layering, etc. Arranged in order according to their proportions are: multi-dimensional vertical base height or horizontal arrangement (22%), solid tower layer (20%), overhead (15%), surrounding or regular surrounding (15%), three-dimensional enclosure half enclosure + top (8.3%), three-dimensional horizontal enclosure (6.7%), three-dimensional structure consisting of a two-dimensional structure with internal space (5%), three-dimensional level surrounded by two blocks height + top + internal space (5%), single-dimensional vertical base height and extension (3%). The results of this experiment show that the level of building blocks for pre-school children aged 4-6 has not yet reached a good level (lower-moderate), and their skills are in the transition stage from a two-dimensional structure to a three-dimensional primary structure.

Table 3.5 Children's building blocks construction level (drawn by the author)

Grade of building block construction	Level of building block construction	Number of people	Percentage of different levels (%)	Percentage of grades(%)
Grade 1 (single-dimensional)	1	2	3.0	3.0
Grade 2 (two-dimensional)	2	14	22.0	52.0
	3	9	15.0	
	4	9	15.0	

continued

Grade of building block construction	Level of building block construction	Number of people	Percentage of different levels (%)	Percentage of grades(%)
Grade 3 (three-dimensional)	5	12	20.0	25.0
	6	3	5.0	
Grade 4 (three-dimensional surrounding)	7	5	8.3	20.0
	8	4	6.7	
	9	3	5.0	

Age and gender differences in the development of building block construction level of 4-6 years old pre-school children: This experiment shows that there is a significant age difference in the average building skills of children aged 4-6. The researcher used univariate analysis of variance, with the average building skill score as the dependent variable, and gender (male, female) and age (4, 5, 6) as the independent variables. The main effect of age is significant, $F(2,55) = 8.705$, $p = 0.001$, $\eta p2 = 0.240$. Six-year-old children are significantly higher than four-year-old children ($p < 0.001$). Four-year-old children are marginally smaller than five-year-old children ($p = 0.092$). As shown in Figure 3.5, as young children grow older, the average score of building skills increases significantly. The construction level of four-year-olds and five-year-olds is mainly concentrated in grade two. There is no one with the highest level of construction among four-year-olds. The construction level of six-year-olds is mainly concentrated in grades three and four, accounting for 40% respectively. It can be seen that the building level of six-year-old children is significantly higher than that of children in other age groups.

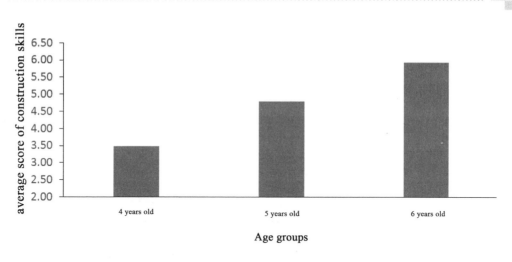

Figure 3.5 Level of building blocks construction of children of different ages
(drawn by the author)

There is no significant difference between the average score of building skills and the gender of young children. As shown in Figure 3.6, there are more boys than girls with higher skill levels, and more girls than boys with lower skill levels. The experimental research of Rogers abroad proposed that the use of enough building blocks, the way of game interaction and suitable auxiliary materials can reduce the gender difference of children in building block games. Both boys and girls can obtain the same building block building time and building block building experience through building block games, which reduces the impact of gender differences to a certain extent.

② **Analysis of Children's Creativity Level:** Experiment II of this research used the TTCT-A graphic drawing assessment tool to evaluate the development ability of 58 4-6 year old children's drawing creativity level, with a maximum score of 22 points. The results of this test show that the average score of 4-6 years old children's painting creativity is 10 points, and the highest level of creativity is 19 points, which indicates that the development of children's painting creativity is at a low-middle level.

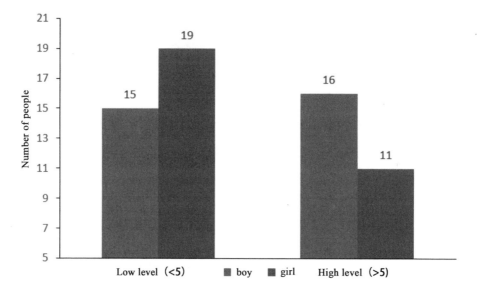

Figure 3.6 Level distribution of building blocks of children of different genders (drawn by the author)

The study uses univariate analysis of variance, the dependent variable is the level of creativity in painting, and gender (male, female) and age (4, 5, 6) are independent variables. The main effect of age is significant, $F(2,53) = 11.863$, $p < 0.001$, $\eta p2 = 0.309$. Six-year-old children were significantly higher than five-year-old ($p = 0.004$) and four-year-old children ($p < 0.001$). The main effect of gender is marginal, $p = 0.067$. Girls have a tendency to score higher in creativity than boys. The main reason is that pre-school girls develop earlier than boys, and girls are more willing to listen to teacher guidance in kindergarten. Their learning attitude and performance are more serious, and their learning knowledge is more solid and in-depth than boys. The influence of gender and age on the level of creativity in painting is shown in Figure 3.7.

In this research, the level of children's drawing creativity is divided into three dimensions, namely fluency, originality, and diligence. Among them, the highest score in the fluency dimension is 15 points, and the lowest score is 2 points. In the diligence dimension, the highest score is 4 points and the lowest score is 1 point, which has a great impact on the overall level of creativity in painting. The development of children's drawing creativity level is greatly affected by fluency and diligence. Among them, the fluency drawing ability is at a low level, and the average diligence is moderately low, which directly leads to the low level of children's overall creativity.

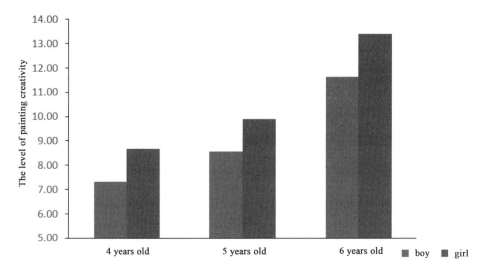

Figure 3.7 The level distribution of the drawing creativity of children with different genders and ages (drawn by the author)

③ **The relationship between children's construction level and creativity level:** Experiment III of this research conducted a correlation analysis between the building block construction level of 58 children and the graphic painting creativity level. There is a positive correlation between construction skills and creativity level. On the one hand, the higher the construction skills, the higher the level of creativity in painting. On the other hand, the method of evaluating the creativity level of painting is applied to the building block works, namely, three dimensions: fluency (the number of building blocks) is composed of 1-10 points, and the number of building blocks within 10 is 1 point, and 1 point is added for each increment of 10 blocks; uniqueness (the form of building blocks) is divided into 0-2 groups, in which 0 score is given for the proportion of the form appearing more than 10%, 1 point is given for the proportion of the form appearing between 5% and 10%, and 2 points are given for the proportion of the form appearing less than 5%; diligence (application of auxiliary materials and floor height) is composed of 1-5 points, 1 point is awarded for single-dimensional building, 1.5 points for adding auxiliary materials, and 2 points for single-dimensional floors higher than 5 floors. 2.5 points for 2D construction, 3 points for adding auxiliary materials, 3.5 points for 2D masonry higher than 5 floors, 4 points for 3D construction, 4.5 points for adding auxiliary materials, 5 points for 3D construction higher than 5 floors. This evaluation tool was used to evaluate the creativity level of building block works, with a maximum score of 17 points, and the highest score of the subjects in this research experiment was 14 points.

The correlation test between the average score of building skills and the level of creativity in painting and the score of building block creativity was carried out. The results showed that the three types of scores were significantly positively correlated.

Chapter 3 User Research in the Interaction Design of Children's Products

Building skills and painting creativity level (r = 0.556, p < 0.001) and building block creativity level (r = 0.735, p < 0.001) are both significantly positively correlated, and there is also a significant positive correlation between the two types of creativity (r = 0.476, p < 0.001), as shown in Figure 3.8 below. From this experiment, two points of view are drawn: ① The TTCT-A graphic drawing evaluation tool can be used to evaluate the level of creativity of building blocks. ② The improvement of the level of building blocks play has a significant positive correlation with the level of creativity of children.

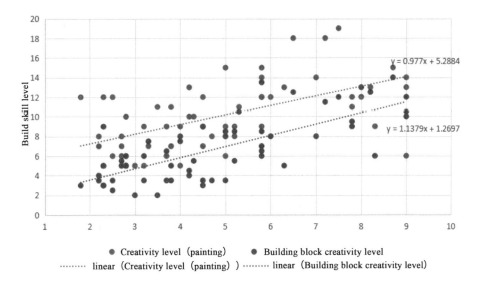

Figure 3.8 The relationship between construction skills and creativity (drawn by the author)

3.2.3 Analysis of children's emotional experience in games

Emotional design research conducted quantitative analysis and design research on children's emotional experience in games from the perspective of design psychology, refined three typical types of game emotional experience, and examined the relationship between the level of game construction and the types of emotional experience. Finally, the principles of emotional experience design for children's constructive puzzle games are proposed, and elements that conform to children's emotional experience mechanisms are injected into product design to accelerate children's immersive game experience.

① **Perceiving children's emotional experience levels in the game process.** The emotional experience in the game is about the process of communication. During the interaction between children and games, the emotions undergo subtle changes, either stable, or ups and downs, or immersed, or free ... According to the complete communication process of the game, the researchers divided the entire emotional experience process into three parts: before the game, during the game, and after the game, and observed the facial expression changes of 61 subjects in the videos. Through the different stages of children's play, the researchers divided the emotional experience of 61 subjects into 6 levels according to the psychological test scale, which are 1 anger, 2 unhappy, 3 concentration, 4 happiness, 5 excitement, 6 expecting. Through video observation, 3 researchers screened out 7 levels of expression evaluation criteria in the scale.

② **Coding of emotional effects during the game.** Emotion is composed of three components: A. Subjective experience (self-feeling). B. External performance (facial expression, posture). C. Physiological response (emotion produces various physiological reactions and different reaction modes). This article mainly studies the explicit emotional experience and implicit emotional experience through external performance (facial expression). Through the game video of 61 children, this study uses the 7-level expression scale in step 1 to encode the emotional changes of the testers during the game process (once every 3 minutes, if there are emotional changes within 3 minutes, the researcher will record the score, time and reason of the changes). Finally, the researchers created 61 graphs of the emotional ups and downs experienced during the game. The researchers analyzed and summarized 61 emotional experience curves, and extracted three typical curves of emotional experience in games, as shown in Figure 3.9. The three types of graphs describe: Active type: emotional fluctuations, instability, and poor tolerance. Concentrated type: long-lasting attention, strong persistence, not easily distracted. Neutral type: not meeting the above two types, having poor adaptability, flexibility, and shyness to the environment and people.

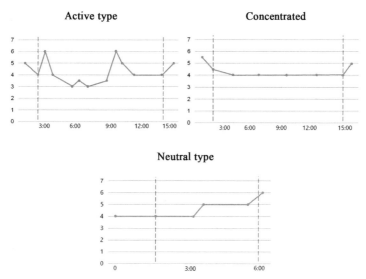

Figure 3.9 Three typical types of emotional experience in the game process
(drawn by the author)

On the one hand, this study uses the building block construction level test tool written by Casey & Andrews to evaluate children's level of building block construction. The experimental results show that the building block construction level of pre-school children is in the transition stage from two-dimensional to three-dimensional primary structure. That is to say, in the process of building blocks, children will use the skills of multi-dimensional vertical masonry or horizontal laying, overhead, enclosure, and solid layering.

On the other hand, this study uses One-Way ANOVA to test whether the different levels of a factor variable will cause significant differences or changes in the dependent variable. In this experimental data, the dependent variable is the average score of building skills, and the independent variable is the emotional type (neutral, active, focused). The test results show that the main effect of the game emotion type is significant, the F value of the test for homogeneity of variance $(2, 58) = 3.269$, and the reliability of the result p value $= 0.045$. The next step is to perform multiple comparison tests. The results show that the building skills of focused children are significantly higher than those of neutral children, with a p-value $= 0.042$ (as shown in Figure 3.10 below). It can be seen that in the game, focused children are conducive to the improvement of building skills. In the product design of building block games, guiding pre-school children to obtain a focused emotional experience in building block games helps to achieve better game effects and stimulate children's interest in games.

③ **Design factor extraction of emotional touch points in games.** Positive emotions can stimulate children's interest in learning and promote the interaction of games; at the same time, appropriate negative emotions are also very helpful to games. On the one hand, they can enhance children's sensory experience in games, and on the other hand, they can also exercise children's ability to face, process, and control negative emotions. Therefore, in the design of children's products, it is necessary to consider how emotional design factors are integrated into the interaction of children's games and applied to the design and development of children's products. In this experiment, what are the design factors associated with the positive and negative touch points of the 61 subjects' emotional changes during the game? In which dimensions are they distributed? Next, the researcher takes the emotional experience curve in step 2 as an example, extracts the touch factors that affect the emotional changes of the game, and establishes the mapping relationship between the touch factors and product design. The process is as follows

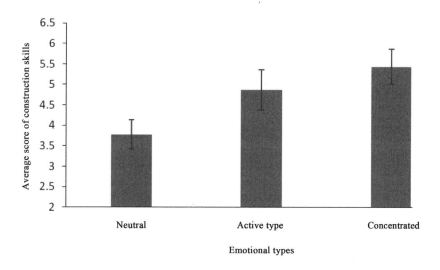

Figure 3.10 The relationship between children of three types of typical game emotional experience and their building skills (drawn by the author)

A. The acquisition of children's primitive subjective emotions. To understand the reasons for children's emotional high and low touch points in the game, first, two researchers extracted relevant information from the video. Because children's emotion judgments are subjective and vague, the initial materials extracted are mostly rough and scattered, and need to be refined many times before they can be transformed into reasonable and pertinent design factors. After the experiment, the author analyzed the emotional change touch points of 61 subjects, and counted a total of 123 original descriptive sentences of the reasons for the high and low touch points of emotional fluctuations. Then, the two researchers conducted semantic analysis on the language of 123 items together, removed items with repetitive semantics, and integrated items with similar semantics. After this step is completed, 50 original semantic items are condensed, and the item set is marked as R, $R=\{ rm \mid m= 1,...,M\}$, $M=50$, and the analyzed data items lay the foundation for the next step.

B. Researchers refined design attractive items from the original semantic items. Because the children's emotional ups and downs are caused by the most primitive emotional reactions and life-oriented language fragments, the 50-item semantic description extracted in the first step is basically subjective, superficial, and concise words and sentences. The 50 simple items need to be translated into objective and scientifically designed attractive items, in order to express the subjective emotional factors more objectively. After finishing the original subjective emotions, the author summarized 50 original semantic items, and converted the 50 translations marked R into 38 attractive design items, marked I, $I=\{ im \mid m=1,\cdots, M \}$, $M=38$.

C. The definition of the dimension of emotional game design. The emotional design of pre-school children's games is a collection of multiple dimensions of design factors. The author will define the content covered by 38 attractive design factors in this round. The second step of the experiment has refined the factors that affect the emotional design of the game. Next, the researcher will apply the factor analysis method to classify the design attractiveness items by dimension, and define the classification and name of each design dimension. According to the statistical results of factor analysis, a total of 38 attractive design items were extracted. According to the distribution of correlation coefficients, they can be divided into six dimensions, each of which contains semantic content that is commonly described. On the other hand, 38 items will be translated into written semantic set W according to semantics, $W=\{ W_a \mid a=1,...,A\}$, $A=38$ in order to describe and define the content of dimensions. In the end, the author summarized the dimensions that affect the emotional design of the game, the related emotional touch items included, and the classification weight (of all touch points that have emotional changes, the frequency of each dimension) as shown in Table 3.6.

Table 3.6   Classification of dimensions that affect the emotional design of block games (drawn by the author)

Design dimension	Emotional touch items	Weight of the classification (%)
The diversity of the appearance of building blocks	w10. The building blocks are not built into tall buildings, which makes oneself unhappy; w14. The building blocks are very high, which makes oneself very excited; w16. Building block materials can build different shapes, which is exciting; w17. The overhead structure built with building blocks is very attractive; w18. The house and the little figures built by building blocks make me very excited; w20. The cute shape with wings built from building blocks makes me excited; w29. Building higher and higher stairs with building blocks makes me happy.	15.2

continued

Design dimension	Emotional touch items	Weight of the classification (%)
The aesthetics of building block design	w6. There is no suitable geometry in the building block, which makes me hesitate to choose; w31. The building block works satisfy me, and the size of the building blocks suits me well; w32. The brick geometry is not suitable, but I am happy to find other shapes to solve the problem; w36. The quick assembly of blocks makes me happy; w38. Looking for suitable building blocks in building block construction makes me look forward to.	14.3
The attraction of auxiliary materials	w19. I look forward to more furniture auxiliary materials used as building block materials; w25. The pony with building block auxiliary materials is impressive, and it can also be inserted with building blocks; w26. The windows with auxiliary materials of building blocks are very attractive; w30. The roof built with auxiliary materials is impressive.	8.1
Sharing of block game content	w21. Building block game interaction makes people very happy; w24. The description of the sharing of building block works makes me very happy; w35. It makes me happy to share what is being built at the moment in the building block construction; w37. In the building block game, I like to play with my friends in the construction area.	22.8
The challenge of building block games	w1. The building blocks suddenly collapsed during the construction, making myself anxious; w2. The building blocks were disassembled and rebuilt, which made me anxious; w3. I don't know how to build the building blocks, which makes me unhappy; w4. The building block materials were difficult to connect, so I asked for help; w7. There is not enough time for the building block game, and the construction process is slow; w27. The process of the building blocks game makes me happy; w33. The collapse of building blocks is interesting.	18.6

continued

Design dimension	Emotional touch items	Weight of the classification (%)
The gradual progress of the block game content	w5. The building block game makes me nervous, and I refuse to communicate; w9. I have lost interest in the building block game; w11. The building block game makes me look forward to it; w12. The building block game makes me feel very fulfilled; w13. The building block game is very attractive; w15. After the end of the building block game, I am very excited; w22. The building block works satisfy me; w23. I am very happy when the building block game is almost finished; w28. The building block works are not finished, but I feel satisfied; w34. The building block game is very fresh.	21.0

3.2.3 Personas based on data analysis

Building personas with scientific methods is an urgent need for the design of children's products. Persona, also known as service role playing (user scenario), originated from one of the methods used by IDEO design company and Stanford University design team to conduct IT product user research. The father of interaction design Alan Cooper first proposed the concept of "persona". In his opinion, by establishing a virtual representative of real users, that is, by "drawing" a virtual user based on a deep understanding of real data, product development or service design can be carried out for the target user group, so as to achieve on-demand mass production and private customization, so as to build enterprise development strategies.

The establishment of a persona is mainly based on the quantitative and qualitative analysis of survey data. The construction method mainly includes 4 steps, that is, collating data, combining character features, refining the persona, and verifying the persona (as shown in Figure 3.11). In the planning stage of this chapter, the researcher first collected the most primitive qualitative data (that is, the most primitive user profile) through in-depth interviews with children of a certain age. Then, the researchers further refined the user portraits obtained through quantitative research methods such as questionnaire surveys and natural observation methods, so as to gradually make the children's persona gradually clear. The specific ideas of persona construction are as follows, which will be described separately below.

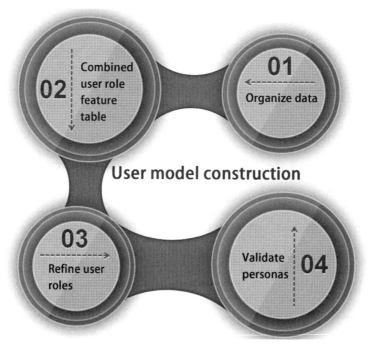

Figure 3.11　Ways to build personas (drawn by the author)

Chapter 3   User Research in the Interaction Design of Children's Products

1. Sorting out data

After the user survey is over, when sorting out the data, first distinguish the child categories. The classification of child user categories is conducive to associating the rational and tedious data features with the user roles in the designer's mind, which is conducive to the summary and extraction of the data, so that the data can be analyzed and summarized to quickly create a persona. The preliminarily determined child category in this survey is 4-6 years old pre-school children, and games are usually the typical way of activities for children at this stage. After roughly clarifying the target population and its basic characteristics, the researcher began to process the data. Researchers use the affinity diagram method to summarize and sort a large number of facts, opinions, ideas and other language materials according to affinity, mainly according to the following stages:

① **Conceptualization of the basic characteristics of user groups.** Human characteristic attributes belong to a combination of cross-dimensions. The cognitive level, perception ability, and intelligence level of pre-school children and adults are completely different, and there is a significant group tendency. Based on the survey data of in-depth interviews, the researcher initially outlined the following characteristics of children in this group:

- In terms of color cognition, pre-school children have basic color discrimination capabilities and are easily attracted by bright colors.
- In terms of mathematical operations, pre-school children have a computational ability within 10-50.

- In terms of figure cognition, pre-school children have the cognitive ability of concrete figures but not the cognitive ability of abstract symbolic representation.
- In terms of sorting and merging, pre-school children have the ability to sort intuitive objects such as height, length, storyline, etc.
- In terms of spatial cognition, pre-school children do not have the cognitive ability space division and three-dimensional construction.
- In terms of attention and memory, pre-school children have strong short-term memory skills, and have a short time of concentration, and are very easy to be distracted.

② **The production of subdivision and induction graphs.** The author turns the collected conceptual information into cards, and invites those involved in data collection to make and discuss the subdivision summary graph. Only one piece of information should be written on each card, and the content includes the objects, behaviors, and problems encountered by children of this age group. Then, the cards are randomly posted on the wall for people to browse to generate new ideas, and then start sorting and classification after the total number reaches 30, as shown in Figure 3.12. After completing several rounds of classification and sorting according to the pre-supposed assumptions, a stable order is basically obtained. That is to say, pre-school children play with toys on average every day, they like to play with good friends, and their play behaviors are mostly based on natural interaction, and toy expenses account for a large part of the money their parents spend for their children.

Figure 3.12 Preliminary classification of induction graphs (drawn by the author)

2. Combining the user characteristics tables

After the data is sorted, the designer usually gets several drafts of the persona. In the comparison of role models, it is inevitable to eliminate some categories with lower utilization value. The appearance of the user role feature table simplifies this process. It is like a product sketch. A rough sketch can be drawn with a few strokes to reduce the output cost of the program.

① **Refining the categories of children's group characteristics.** The researchers made a general segmentation of children in this age group, and differentiated and compared the content in the age grouping of 4-6 years old pre-school children. This book uses questionnaire surveys to draw a more complete portrait of the population in terms of game behavior attributes, children's product preference categories, brands, product features, and sensory preferences among 109 children aged 3-7 years old. It has a deep understanding of the target population from the aspects of physical development, psychological development, cognitive development, and game development to grasp their core needs.

② **Creating the persona feature table.** The designer constructed a persona feature table for children of this age, and extracted the main tags after data analysis to form a list of basic information for this persona. The persona feature table of each age group should be added with close to the real role information, and the persona information should be as full as possible, as shown in Table 3.7.

Table 3.7 Characteristic table of children personas (drawn by the author)

Age	Physical development	Psychological development	Cognitive development	Game development
4 years old	Average height 110.5 cm; sitting position 91.8 cm; arm length 40.8; palm length and width 10.7/5.7 cm	Low concentration in game emotional experience; the level of creativity in painting and construction is low, and the two types of creativity are significantly positively correlated	In terms of color, they prefer purple, green, and yellow; in terms of graphics, they like guns, cats and dogs, and flowers; in terms of materials, they like plastic products	Game building skills are mainly low-level single-dimensional building; they gradually start to develop reading habits, prefer building blocks, hide and seek

continued

Age	Physical development	Psychological development	Cognitive development	Game development
5 years old	Average height 118.3 cm; sitting position 93 cm; arm length 43.9; palm length and width 11.2/5.9 cm	Their concentration and activity in the emotional experience of games is higher than that of four-year-old children; their level of creativity in painting and construction is low, and the two types of creativity are significantly positively correlated	In terms of color, they prefer red, yellow, and blue; in terms of graphics, they like flowers, cars, cats and dogs; in terms of materials, they like plush and metal products	Their construction skills began to transition to two-dimensional overhead and surrounding; various game themes are involved, and they prefer block games
6 years old	Average height 122.5 cm; sitting position 95.7 cm; arm length 45; palm length and width 11.8/6.1 cm	Their emotional experience of games is the most active; their painting and construction creativity levels are significantly higher, and the two types of creativity are significantly positively correlated	In terms of colors, they prefer blue, green, red, and purple; in terms of graphics, they like guns, cars, and geometric shapes; in terms of materials, they like plastic and metal products.	Their construction skills are at the two-dimensional level of enclosure and three-dimensional level enclosure; their preference for video games has increased significantly, and they have a preference for building block games

3. Refining the personas

After the user feature list is completed, these character sketches can be developed into formal personas. The personas should include some key information that can be used for definition: goals, roles, behaviors, environment, typical activities, etc. These contents make the personas richer, fuller, closer to the real situation, and more importantly, they are closely related to the product. ① The name of the user role. Characters without names are cold and digitized. On the one hand, names reduce the designer's memory burden, and on the other hand, they can also serve the generalization of labeling. In order to clarify the differences between pre-schoolers, we can briefly describe the names of different age characteristics. For example, we can use "the shy Zhang Zihao" and "the perfectionist Wang Lele" to summarize the roles of 4 and 6-year-old children to make the roles easier to identify. ② Selection of persona photos. First of all, use real children's photos, and absolutely not use comics and gallery avatars; second, as far as possible in terms of the diversity of photo details, show photos with a uniform style and shooting techniques as much as possible. ③ The establishment of user documentation. We can use a one-page user document as a bridge between the role sketch and the plump persona. The basic information of the persona is as shown in Table 3.8.

Table 3.8 Basic information of preschooler personas (drawn by the author)

Goal	Demographic features	Labels	Skills and knowledge	Environment
Building block game, theme building	4 years old, boy, Zhang Zihao	Shy, poor concentration when playing games, not very good at communicating with others	He often plays blocks and hide-and-seek games, and likes to build cars with plastic blocks, but he always doesn't put them together well.	During the play time in the kindergarten, he likes to build by himself in the construction area, not looking very gregarious
Cooperate with friends for themed block building	6 years old, boy, Wang Lele	Perfectionism, focused when playing games, likes to share the fun of games and competition interaction with friends	He often plays building blocks and video games, and uses blocks of different materials to build a theme park. He has high game skills and can build a three-dimensional castle.	He invites children to play with building blocks at home, takes out his favorite fire station series, and tries to challenge difficult buildings

Through the specific and enriched user model feature table, we use the original data description in the user model description document as much as possible. After the above steps, two user models are generated. The main persona is Wang Lele (playing interactively with children), and the secondary persona is Zhang Zihao (playing alone), as shown in Figure 3.13.

Main user model
Information generalization **Name:** Wang lele **Age:** 6 years old Playing with building blocks: about 3 years **Daily game time:** about 4 hours **Game location:** home living room **Game partner:** neighbor Ding Ding **Main problems:** toys are tired, not attractive and interactive **Personal description** In the red, yellow and blue kindergarten, he will go to primary school soon. Perfectionism, like to share the fun of games with friends and participate in competitive activities. He often plays building blocks and video games. He has a high level of building blocks, can build three-dimensional castles, and is eager to try to challenge difficult building collages. **Problem description** At present, the traditional building block toys on the market have been played all over, and we feel boring and have no interest. And friends are eager for interactive building block toys and want to compete together. The difficulty of building blocks challenge at home is too low. I feel that there is no difficulty and I am eager to try more. When playing games, I am fascinated and often forget to eat and rest. There are too few auxiliary materials for building blocks, and there are no more building block elements with different shapes.

Secondary user model
Information generalization **Name:** Zhang Zihao **Age:** 4 years old Playing with building blocks: about 0.5 years Daily game time: about 6 hours Game location: kindergarten Game partner: teachers, parents **Main problems:** I can't imagine building blocks by myself and don't like to communicate with others **Personal description** In the small class of red, yellow and blue kindergarten, he was shy and didn't communicate with people soon after entering the kindergarten. He often played building blocks in the construction area of the kindergarten and hide and seek games with his parents at home. He liked to build cars with plastic building blocks, but he always couldn't match well. **Problem description** He likes to play with building block toys, but don't know how to spell them. Most of them can only spell according to the drawings. When playing with toys, he don't pay enough attention and is easy to be affected by the surrounding sounds. There are few building block toys in the kindergarten, and the shape of splicing is relatively single. When playing children's games in kindergarten, he doesn't fit in with the crowd and likes to play alone in the construction area. The skill of building block splicing is very low.He hates that the building blocks collapse and cannot be spliced and connected.

Figure 3.13   Persona documents (drawn by the author)

4. Verifying the personas

After the researcher has produced a complete set of personas, each persona is full of various information, including primary and secondary data, quantitative and qualitative data, new and old data. They may conform to each other, or they may be mutually exclusive, so it is very important to verify the completed personas. The purpose of verifying the user model is to ensure that the output user model (virtual story) is consistent with the real data, especially the essence of the target child group must be effectively presented in the data of the personas.

Model verification is mainly done through focus group interviews. The researcher returned to the red, yellow and blue kindergarten and invited 4-6 year-old children who matched the personas to participate in the building block cooperative game again. The researcher invited a total of 12 groups (2 in each group) of pre-school children. In each group, a target user invited one of his friends to participate. The researcher asked them to freely fight with their peers for 15 minutes. Through natural observation and question and answer, the researcher directly obtained and recorded children's feedback, thereby further understanding and improving the user model, and laying a solid foundation for the user-centered design in the next step.

3.3 Kidsplay, a research data platform for children's development

Based on the large amount of literature data collected in the previous period and in-depth user research data, the researcher believes that it is very meaningful to pay attention to the continuous research of children's physical, cognitive and psychological development. This chapter describes in detail the establishment of a data platform for child development research—"Kidsplay". By tracing its growth process, we can understand how "user research" penetrates into each link. "Kidsplay" is a platform based on the analysis of children development research data. Through this platform, we can know the characteristics of children's physical development, mental development, cognitive development, and game development at different ages. Here, users can fully understand the children's design data required by industries such as the design of children's products, education research, and game applications, which makes it a professional children development research platform.

3.3.1 Positioning and mission of the platform

1. The positioning of the platform.

In the early stages of product design, positioning is very important. A clear positioning is not only an analysis of the target product, but also a more detailed description of the usage scenarios of the product, the user's behavior characteristics, and the mission of the product. "Kidsplay" is positioned as a data platform for children's development research. The target users are data lovers, design teams of children's industry companies, and marketing departments of brand owners. The website of the general version is: http://childreninchina.com.

2. The mission of the platform.

The product mission of "Kidsplay" is to deliver the value of child development research data analysis, and to play the value of Internet data in the traditional children's industry. For example, from the collection of children's cognitive data, analysis and statistics of children's characteristics, dissemination and application in related industries, planning and design of related programs and other links. The current research is related to data collection of pre-school children and school-age children. In the future, "Kidsplay" will continue to improve and become a platform that includes data on Chinese children of all ages. Through free, open, and rigorous data analysis, the value of data will be realized and real social significance will be generated.

There are two main usage scenarios of "Kidsplay": One is the children's industry related people who are interested in data, who can dig out some valuable content for the design of children's products on this platform. For example, designers can design smart watches for pre-school children through hand data, color preferences, and graphics recognition data on the physical development of children in pre-school age. Teachers at the Maker Children's Training Center can learn about school-age children's creativity, emotional experience and other psychological data and game skill development data to design Lego programming courses for school-age children. On the other hand, when studying child development, professionals in children psychology and sociology can learn about the living conditions, characteristics, preferences and needs of children of relevant ages, and provide reference data for their academic research. At the same time, they can upload relevant experimental data to the platform to further improve the research results of the data platform and serve as a bridge for academic exchanges.

3.3.2 Birth of the platform

The birth of the "Kidsplay" platform is the same as other platforms relying on the Internet. There are many pre-design activities, such as establishing platform goals, analyzing the needs of the target population and organizational structure, improving data content, designing interfaces, etc. The American scholar Jesse James Garrett (2003) divided the process of web development experience design into five levels, from bottom to top they are strategy level, scope level, structure level, skeleton level, and surface level, as shown in Figure 3.14.

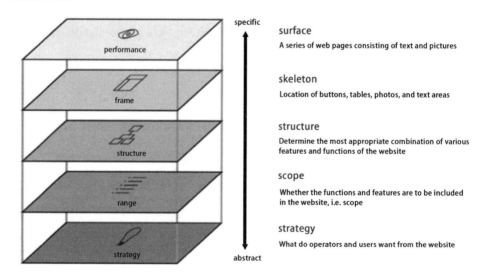

Figure 3.14　Elements of user experience[①]

---

① [美] Jesse James Garrett. 用户体验要素. 范晓燕, 译. 北京: 机械工业出版社, 2011.
   [US] Jesse James Garrett. User Experience Elements. Translated by Fan Xiaoyan. Beijing: China Machinery Industry Press, 2011.

Chapter 3   User Research in the Interaction Design of Children's Products

User research is integrated into each stage of product development, from the initial strategic level demand collection to the scope level functional definition, from the design of the interactive information architecture to the visual design of the specific interface. In every link, the lack of user research may cause product functions to deviate from user needs, which will lead to huge rework during platform iterations. If the problem is discovered after the platform is online, it will bring huge losses. This will not only waste a lot of design and development resources, but also consume the trust of users, who will most likely never come back. Iterative research is the key to ensuring product survival. Now, the author analyzes the process of how the platform was born as follows:

1. Strategic level design

The target users of the "Kidsplay" children's data platform are data enthusiasts, children's industry design teams, and the marketing department of brand owners. The core demand of users is to understand the characteristics of various aspects of children development research through the data platform, and to play the value of Internet big data for the traditional children's industry product design. The target users of "Kidsplay" are relatively small, so you can conduct in-depth interviews with the target users. Even the author belongs to the core group of target users.

2. Design of the scope level

"Kidsplay" is a newer data platform website. The core function of the product is to present research data on children development (currently it mainly collects some types of research data form preschool children and school-age children, and will continue to be expanded and updated in the future). The second is to provide solutions for children's products based on data research. At present, it mainly focuses on design fields such as entertainment and education (shown in detail in Chapter 6). Finally, the basic information of the platform (such as homepage, about us, display of relevant survey questionnaires, etc.). The home page of the platform includes regular sections such as navigation, key columns, important news, and login and registration. When designing key columns, we ensure that the data collected is clear. We expect that more than half of the visitors will focus on "key columns" and 20% will focus on "navigation". The rest are hybrids. Users who focus on key columns are task-centric and desire to find what they want.

3. Design of the structure level

The priority order of the functions of the "Kidsplay" data platform is already very clear. It's main process is that users search for relevant data through four quick entries in the homepage navigation directory or important columns, and use "solutions" to understand the design of children's products based on data research.

4. Design of the skeleton level

The most important thing about the "Kidsplay" platform is the homepage design, which is the core hub of the entire data platform, and the top priority of the homepage is the "important column" of children development research. Therefore, we need to put the four different types of data entry, the most important "important column", at the

Chapter 3   User Research in the Interaction Design of Children's Products

core. The purpose of such typesetting is to allow users to pay attention to the core information content at a glance. In addition, the navigation directory, as the first-level content of the information architecture, should also be placed in an easy-to-read area on the interface.

5. Design of the surface level

The visual design of "Kidsplay" mainly considers three aspects: ① Consistency of interface design. We position the platform style as a cute and concise children's website. Banner adopts a children's graffiti-like background design to strengthen the theme, and the menu bar and other entrances use icon design language, which is easy to read and recognize, and is also in line with the overall style positioning. ② Contrast design of important information. In the four categories of "important column" data on the homepage of the platform, we "differentiate" the content of other elements on the page through icon design with exaggerated size and sufficient resolution. In the statistical analysis report on the secondary page, we use the information design of a large number of statistical charts to contrast with the text content, which is more accessible and easy to read and convenient for visitors to understand intuitively. ③ Standardized color matching and typesetting. The overall color scheme of the platform is tender green #4AB301 (key tone) and orange # FFD02E (color matching) so as to convey the distinctive style of children's products through bright and eye-catching colors. The text is uniformly black, the main body font is Microsoft Yahei, the logo and title fonts are Tengxiang BRIC black simple, and the English font is uniformly Berlin

Sans FB. In terms of typesetting, the homepage adopts a simple skeletal typesetting. By dividing the larger size of the banner and important column screens, the relationship between the various parts of the homepage is clarified, showing good contrast, making the overall screen not monotonous and crowded.

3.3.3 Launching of the platform

After the interaction details and visual design were completed, the researcher conducted a website development test. The product has been officially launched. Users can browse the relevant data through the website: www. childreninchina. com. If you need to download and view all the documents, users can add a public WeChat account on the homepage to apply for a user name and password. After the product goes online, it is necessary to establish user feedback channels, maintain user groups, monitor background data, and record user behaviors. The picture below shows the final product plan among the countless versions of "Kidsplay". The visual elements of this version of the design are very rich, with information visualization elements such as histograms, area charts, and bar charts. At the same time, the content of each column is organized and summarized, so as to clearly and effectively convey the research data. In the later period, we will continue to improve and update various research data, and establish a WeChat group to invite website designers, engineers, and scientific researchers to join the group. Through children's interactive experimental research, people can more deeply feel the needs of users, and provide better solutions for children's education and product design. At the same time, we will establish a user feedback platform on the data platform. Through this channel, people can understand the problems of platform viewers and collect more valuable needs. Website administrators can respond to user questions through this platform, and establish contact and interaction with users. Finally, through this platform, people can also recruit volunteers, and even conduct small user research directly in the user groups.

3.4 Summary

This chapter is a study of children's users based on the interaction of physical products. It integrates research methods such as interview method and experimental observation method to understand children in an all-round way and provide a theoretical basis for the next step of interaction design of children's products. Firstly, through user interviews and questionnaire surveys, this book obtains first-hand information about children's product preferences and gaming behaviors. Secondly, for the experimental observation method of constructing games, this chapter refines deeper data such as children's game emotional experience characteristics, creativity level, game skill level, physical development characteristics, etc. It summarizes the data analysis results of children's physical development (average height, sitting posture, hand size) and psychological development (emotional experience, creativity level, construction and creativity level correlation), cognitive development (color, graphics, texture recognition), and game development (game skill level, game theme).

The role of this chapter is to carry on the results of the theoretical analysis of Chapter 2, and contribute to the core results of user research—the Kidsplay children development research data platform. It provides effective data products for the children's design ecosystem, conveys the value of user research data, and also brings out the value of Internet data for the traditional children's industry.

References

Original works in foreign languages

[1] Kristina Andersen. "ensemble": Playing with Sensors and Sound[M]. In CHI 2004 Extended Abstracts on Human Factors in Computing Systems, ACM Press, 2004:1239-1242.

[2] McNerney. From turtles to Tangible Programming Bricks: explorations in physical language design[J]. Personal Ubiquitous Computing, 2004, 5(8):326-337.

[3] Bohn, Jurgen. The Smart Jigsaw Puzzle Assistant: Using RFID Technology for Building Augmented Real-World Games[J]. Workshop on Gaming Applications in Pervasive Computing Environments at Pervasive, 2004.

[4] H Raffle, L Yip, H Ishii. Robo Topobo: improvisational performance with robotic toys[J]. Acm Siggraph Sketches, 2006, 1(6): 140.

[5] E Schweikardt, MD Gross. roBlocks: a robotic construction kit for mathematics science education[J]. International Conference on Multimodal Interfaces, 2006: 72-75.

[6] Dan O'Sullivan, Tom Igoe. Physical Computing[M]. Thomson Press, 2004.

[7] Markus Funk, O Korn, A Schmidt. An Augmented Workplace for Enabling User-Defined Tangibles[J]. Extended Abstracts of the Acm Conference on Human Factors in Computing Systems, 2014 :1285-1290.

[8] B Ullmer, H Ishii. Emerging frameworks for tangible user interface[J]. Ibm System Journal, 2000, 39(3.4): 915-931.

[9] D Merrill, J Kalanithi, P Maes. Siftables: towards sensor network user interfaces[J]. International Conference on Tangible & Embedded Interaction, 2007:75-78.

[10] George W. Fitzmaurice. Graspable User Interface[D]. University of Toronto, Toronto, 1996.

[11] GW Fitzmaurice, H Ishii, WAS Buxton. Bricks: Laying the foundations for graspable user interface[J]. Sigchi Conference on Human Factors in Computing systems, 1995, 15 (August 10) :442-449 .

[12] Hiroshi Ishii, Brygg Ullmer. Tangible Bits: Towards Seamless Interfaces between People, Bits and Atoms[J]. Acm Sigchi Conference on Human Factors in Computing Systems, 1997: 234-241.

[13] RK Fergus, Hiroshi. Hiroshi Ishii[M]. Spell Press, 2012.

[14] Brygg Ullmer, Hiroshi Ishii. Emerging Frameworks for Tangible User Interfaces[J]. Ibm Systems Journal, 2000, 39 (3.4) :915-931.

[15] Brygg Ullmer, Hiroshi Ishii. The metaDESK: models and prototypes for tangible user interfaces[J]. Acm Symposium on User Interface Software & Technology, 1997: 223-232.

[16] SAG Wensveen. A Tangibility Approach to Affective Interaction[J]. Design, Engineering & Production, 2005.

[17] L Jones. Thermal Touch[J]. Springer. Scholarpedia, 2009, 4(5).

[18] Janice J Beaty. Observing Development of the Young Child[M]. Pearson Education US Press, 2008.